"Nikon" fluor. mic.
"Turner" spectrofluorometer
Xenonlamp

TECHNIQUES IN VISIBLE AND
ULTRAVIOLET SPECTROMETRY

VOLUME TWO
STANDARDS IN FLUORESCENCE SPECTROMETRY

IN THE SAME SERIES

Standards in Absorption Spectrometry
  UV Spectrometry Group
  edited by A. Knowles and C. Burgess

Standards in Fluorescence Spectrometry

ULTRAVIOLET SPECTROMETRY GROUP

Edited by

J.N. MILLER
Department of Chemistry
Loughborough University of Technology

LONDON NEW YORK
CHAPMAN AND HALL

First published 1981 by
Chapman and Hall Ltd
11 New Fetter Lane, London EC4P 4EE

Published in the USA by
Chapman and Hall
in association with Methuen, Inc.
733 Third Avenue, New York, NY 10017

© 1981 UV Spectrometry Group

Typeset by Tek-Art, Croydon, Surrey
and printed in Great Britain at the University Press, Cambridge

ISBN 0 412 22500 X

All rights reserved. No part of this book may be reprinted, or reproduced, or utilized in any form or by any electronic, mechanical or other means, now known or hereafter invented, including photocopying and recording, or in any information storage and retrieval system, without permission in writing from the publisher.

British Library Cataloguing in Publication Data

Standards in fluorescence spectrometry. —
   (Techniques in visible and ultraviolet spectrometry; Vol.2)
   1. Fluorescence spectroscopy
   I. Miller, J. N.   II. Ultraviolet Spectrometry Group   III. Series
   535.8'4   QC454.F6/
   ISBN 0-412-22500-X

Contents

Preface		vii
Membership of the working party		ix
1	**General considerations on fluorescence spectrometry** *J. N. Miller*	1
	1.1 Introduction	1
	1.2 Molecular photochemistry	2
	1.3 Fluorescence instrumentation	3
	1.4 Good spectroscopic practice	4
	1.5 Fluorescence intensities	5
	1.6 Nomenclature in fluorescence spectrometry	7
2	**Monochromator wavelength calibration** *J. N. Miller*	8
	2.1 Introduction	8
	2.2 Characteristics of calibration methods	9
	2.3 Use of spectral lines from the spectrometer light source	10
	2.4 Use of an auxiliary light source	12
	2.5 Use of narrow bandwidth fluorescence maxima of inorganic and organic solutes	12
	2.6 Conclusions and recommendations	14
3	**Stray light in fluorescence spectrometers** *J.N. Miller*	15
	3.1 Origins of stray light and resultant errors	15
	3.2 Stray light in grating monochromators	16
	3.3 Summary and recommendations	18

vi *Standards in fluorescence spectrometry*

4 Criteria for fluorescence spectrometer sensitivity J. N. Miller 20
    4.1 Background: inter-instrument comparisons 20
    4.2 The limit of detection method 22
    4.3 The signal-to-noise ratio method 23
    4.4 Summary and recommendations 25

5 Inner filter effects, sample cells and their geometry in fluorescence spectrometry J. B. F. Lloyd 27
    5.1 Inner filter effects 27
    5.2 Sample cells 38
    5.3 Recommendations 41

6 Temperature effects and photodecomposition in fluorescence spectrometry M. A. West 44
    6.1 Errors caused by temperature effects 44
    6.2 Countermeasures and recommendations for temperature effects 46
    6.3 Errors caused by photolysis effects 46
    6.4 Countermeasures and recommendations 46

7 Correction of excitation and emission spectra G. C. K. Roberts 49
    7.1 Introduction: the need for correction procedures 49
    7.2 Excitation spectra 49
    7.3 Emission spectra 54
    7.4 Polarization effects 62
    7.5 Recommendations 64

8 The determination of quantum yields J. W. Bridges 68
    8.1 Introduction 68
    8.2 Primary methods of determining quantum yields 71
    8.3 Secondary methods of determining quantum yields: use of fluorescence standards 75
    8.4 Other methods of determining quantum yields 76
    8.5 Summary and recommendations 77

    Appendix Corrected excitation and emission spectra 79

    Index 114

Preface

The Photoelectric Spectrometry Group was formed in July 1948 in Cambridge. The Group was born out of a need for a forum of users to discuss problems and methodology associated with the new era of photoelectric spectrometers. Over the years the aims and objectives of the Group have been broadened to include many aspects of ultraviolet and visible spectrometry. In 1973, the Group renamed itself the UV Spectrometry Group (UVSG). The techniques of fluorescence, diffuse reflectance, ORD and CD were included in the Group's interest. In 1979, the UVSG became a registered charity. The present Group membership is some 200 practising spectroscopists, mostly from the UK with a small but growing overseas membership.

In August 1977, the UVSG Committee set up three Working Parties: Cells for UV-Visible Spectrophotometers; Photometric and Wavelength Standards; and the Calibration of Fluorimeters. It was felt that a wealth of information and expertise in the practice of spectrometry was available within the Group and that it was appropriate for this to be gathered together in the form of a number of monographs. Initially the intention was that these should be circulated only amongst the Group membership. However, the suggestion was made that these monographs would be of interest to other scientists outside our specialist Group. The conclusions of the first two Working Parties were combined in Volume 1 of this series, and this monograph summarizes the work of the third Working Party. In both cases, proceeds from the sales will contribute to the UVSG funds, and hence to further work in these important fields.

Throughout its investigations the Working Party has borne in mind the needs of the chemist or biochemist using fluorescence as an everyday tool; where alternative approaches to a problem have been considered, their ease and convenience have been regarded as important factors in their assessment. Similarly a number of topics of more specialized interest have been omitted. These include a comprehensive study of polarization effects and methods of obtaining polarized luminescence data; standards for phosphorescence measurements and for fluorescence and phosphorescence lifetimes; and standards for (and methods of purification of) suitable solvents for luminescence studies. These topics will, it is hoped, be dealt with in the Working Party's second report.

The Working Party's provisional conclusions were presented to a meeting of the Ultra-Violet Spectrometry Group held at the Royal Institution, London in May 1979. A number of comments and suggestions made at that meeting and subsequently have been incorporated into the text. In this connection we are particularly grateful for the comments and information provided by Ms R. Burns of the National Physical Laboratory, and Dr R. E. Dale of Christie Hospital, Manchester.

It is a pleasure to acknowledge also the enthusiastic help and support of Dr M. A. Russell, the UVSG Chairman, who has helped all the Working Parties substantially: of the other members of the UVSG Committee, particularly Drs Burgess and Knowles, editors of the companion volume; of Mr C. S. Lim and Miss H. M. Thakrar for help with the spectra in the Appendix; and particularly of Mary Emerson of Chapman and Hall, our ever-patient publishers.

When the Fluorescence Working Party's second report is published, it is hoped to include a chapter updating and modifying the conclusions of this monograph. Comments or criticisms will thus be welcome.

January 1981   J.N. MILLER

Membership of the working party

Professor J. W. Bridges, Robens Institute of Industrial and Environmental Health and Safety, University of Surrey

Dr J. B. F. Lloyd, Home Office Forensic Science Laboratory, Birmingham

Professor J. N. Miller (Chairman), Department of Chemistry, Loughborough University of Technology

Dr G. C. K. Roberts, National Institute for Medical Research, Mill Hill, London

Dr M. A. West, Assistant Director, The Royal Institution, London

1 General considerations on fluorescence spectrometry

1.1 Introduction

Fluorescence spectrometry is becoming increasingly popular in many branches of the chemical and biological sciences. The rapid growth in the number of research publications, and the frequent introduction of new commercially-available instruments, both testify to the importance of the technique. It is used in studies of molecular structures and molecular interactions, in the localization of molecules (especially in biological systems), and in many types of trace analysis. The principal advantages of the technique which encourage its use in all these fields are: (a) sensitivity; picogram quantities of luminescent materials can frequently be studied, (b) selectivity, deriving in part from the two characteristic wavelengths (excitation, fluorescence) of each compound, and (c) the variety of sampling methods available: dilute and concentrated solutions, suspensions and solid surfaces can all be readily studied, and combinations of fluorescence spectroscopy and chromatography are especially useful.

In spite of these advantages, many laboratories may be deterred from using fluorescence methods by the lack of standardization in the field. For example, most commercially-available instruments generate spectra that reflect the characteristics of the instrument as well as the properties of the sample. Comparisons between results obtained using different instruments are therefore difficult, though several correction procedures are available. Other areas where standards are unavailable or unsatisfactory include the provision of a primary standard for quantum yield determinations, the specifications of sample cells, the stray-light properties of instruments, and a suitable definition of the sensitivity of a fluorescence spectrometer.

These and some related problems are discussed in the present monograph. This chapter is not an attempt to provide a rigorous introduction to fluorescence spectrometry, but rather to present an overview of the subject, highlighting the topics that are studied in detail in subsequent chapters. It should be noted that the contents of the first volume of this series, in particular the chapters on cell design, cell handling and stray light, are of considerable relevance to workers in the fluorescence field [1].

1.2 Molecular photochemistry

The essence of fluorescence spectrometry is that a molecular sample, illuminated by light from an external source, emits fluorescence (and possibly also phosphorescence and delayed fluorescence) at a different wavelength, generally longer than that of the exciting light. The process of excitation takes only ca. 10^{-15} s and is thus effectively instantaneous: however, the excited molecule can lose its excess energy by a variety of mechanisms, whose different rate constants determine the most important deactivation routes. The lifetime, τ_f, of the fluorescence transition, normally a radiative transition from the first excited singlet state S_1 to the ground state S_0 (i.e. $S_1 \rightarrow S_0$), is ca. 10 ns. τ_f is defined as the time taken for the number of molecules in S_1 to fall to 1/e (0.368) of its initial value via the fluorescence pathway: it is thus the reciprocal of the fluorescence rate constant, k_f, in the absence of radiationless deactivation processes. In practice, of course, many of the excited molecules may indeed be deactivated by processes other than fluorescence, and it is possible to define the quantum yield ϕ_f as the fraction of photo-excited molecules to lose their excess energy via the fluorescence mechanism. The determination of ϕ_f values is discussed in Chapter 8.

The nanosecond time interval between excitation and fluorescence is both advantageous and disadvantageous in fluorimetry — advantageous because it can be used to study molecular processes that occur in similar or shorter time intervals; disadvantageous in that it renders fluorescence phenomena vulnerable to many environmental influences. Thus ϕ_f values, in addition to being strongly dependent on structural factors, also depend on factors such as temperature, viscosity, solvent composition, etc., to a far greater degree than do absorption phenomena.

An important competitor of fluorescence as part of a deactivating route is intersystem crossing to the lowest triplet state T_1, a process whose rate constant (k_{isc}) is frequently 10^7–10^8 s^{-1}, i.e. similar to k_f. The intersystem crossing S_1--→T_1) transition* is forbidden in quantum mechanical terms, but in practice can often occur because of state mixing and indeed in some cases may be artificially enhanced by the addition of 'heavy' atoms or ions (Br, I, Cs^+, Ag^+, etc.). A molecule in the triplet state may emit radiation in the form of phosphorescence ($T_1 \rightarrow S_0$). Phosphorescence has a much longer lifetime than fluorescence ($\tau_p = 10^{-3}$–10^2 s) and is thus generally quenched in solution at room temperature by oxygen and other molecules.

Loss of energy by vibrational relaxation and internal conversion involving excited states is generally much more rapid than the fluorescence and phosphorescence transitions, which normally originate from the ground vibrational states of S_1 and T_1, respectively. This energy loss ensures that the wavelength of maximum fluorescence is longer than the wavelength of maximum absorption (solvent relaxation effects also contribute to this result): further, since the T_1 level is of lower energy than S_1, the phosphorescence maximum is at a longer wavelength than the fluorescence maximum.

1.3 Fluorescence instrumentation

From this brief survey of molecular phenomena it is apparent that a fluorescence spectrometer must be capable of measuring light intensities (normally relative intensities, but see Chapters 7 and 8) at various wavelengths from samples that can also be irradiated at selected wavelengths. Light from the source that is not absorbed by the sample (i.e. is transmitted or scattered) is not generally required. The normal instrument layout thus comprises (see Chapter 2) two monochromators, the fluorescence being observed at right-angles to the incident beam. The 90° optics are designed to eliminate transmitted light and to minimize Rayleigh and Raman scattered light interference. In practice these scattering signals (from the solvent) are observed, and Raman scattering may indeed be put to good use in assessing spectrometer sensitivity (Chapter 4). This optical design presents the user with a number of problems. Firstly, both monochromators must be maintained in proper adjustment (Chapter 2). Secondly since (see Section 1.4 below) fluorescence intensity is dependent on incident light intensity, intense light sources

*the broken arrow (- - →) indicates a radiationless transition

may be used, giving rise to both heating and photodecomposition effects (Chapter 6). Instrument geometry also has a profound influence on inner filter effects (Chapter 5). Perhaps most importantly of all, the conventional single-beam optical design gives rise to the curious situation in which excitation and fluorescence spectra apparently vary from one instrument to another: observed excitation spectra are affected by the wavelength-dependent properties of the primary light source and excitation optics, and fluorescence spectra depend on the wavelength dependence of the detector sensitivity and the efficiency of the other emission optics. The 'correction' of such spectra, so that the results of all workers are comparable, is obviously important (Chapter 7).

The previous generation of fluorescence spectrometers were often very simple, incorporating a line source (mercury lamp) and filters rather than monochromators. Some of the standardization problems are absent in such cases (no monochromator calibration, spectra not obtainable, etc.), but other problems (e.g. temperature effects, photodecomposition) may remain. Modern instruments, with the optical layout outlined above, are generally designed for use in the 200–700 nm region. In practice stray-light effects (Chapter 3), absorption of light by atmospheric oxygen (the facility for nitrogen purging of both spectrometer and samples is certainly valuable), and low source output make measurements below ca. 240 nm difficult. The limitations on solvent [1] are also at least as great in fluorimetry as in absorptiometry. At the high-wavelength end of the range, special red-sensitive photomultipliers are necessary for studies above ca. 550 nm. In practice, the region 250–550 nm encompasses the properties of most molecules of everyday interest.

1.4 Good spectroscopic practice

The extreme sensitivity of fluorescence spectrometry demands a very high standard of experimental work, if optimum results are to be achieved. Like most techniques it also involves certain hazards. A brief summary of desirable precautions therefore follows.

(a) The principle hazards, apart from those normally associated with electrical equipment, are those involving the high pressure xenon lamps still widely used as light sources. These lamps contain gas at several atmospheres pressure and thus should be handled with great circumspection: eye protection, heavy gloves and a chest protector are recommended, and in no circumstances should the lamp be touched when hot. Occasionally, the lamps fail while in use. This results in a loud report and the production of numerous tiny shards of glass or silica; so the lamp should always be surrounded by a stout metal casing. In many instruments the lamps are cooled by a fan; if the fan fails, the lamp should be switched off at once. Other hazards associated with xenon lamps include ultra-violet radiation – the lamps should never be observed directly or by reflected light without the use of eye protection – and the emission of ozone. (The use of ozone-free lamp-houses is becoming more common and is much to be encouraged).

(b) Fluorescence spectrometers should be used in dust-free laboratories with small temperature variations. While this may not always be feasible in practice, it is certain that using an instrument in a laboratory where 'wet chemistry' experiments are carried out can cause rapid deterioration of the optical components, especially those near the hot light source. In any case mirrors should be carefully cleaned at regular intervals.

(c) Extreme precautions must be observed with regard to the cleanliness of cells. Fingerprints exhibit substantial fluorescence [2] so the handling of cells must be carried out with care and their faces wiped with lint-free lens tissues (not conventional laboratory tissues). Cells must be cleaned in fairly rigorous conditions, as some solutes (e.g. proteins) adsorb strongly to their surfaces – but the cleansing agent must also be non-fluorescent. A recent striking example of this type of contamination was the demonstration [3] that the well-known. Decon detergents react with the fluorigenic reagent o-phthalaldehyde.

(d) Similarly, samples should be stored in clean glass vessels (not plastic containers or glass vessels with plastic stoppers) and should normally be free of dust and suspended particles. A convenient technique involves passing the sample through a 0.22 μm membrane filter – this will also remove bacteria which may interfere with certain fluorescence analyses [4].

These precautions are, of course, additional to those described in subsequent chapters of this book.

1.5 Fluorescence intensities

The relationship between the fluorescence intensity of a dilute sample and its concentration is described in detail in many texts, e.g. [5], but can be simply derived from the Beer-Lambert Law of absorption as follows:

The Beer-Lambert Law states that

$$I_t = I_0 10^{-\epsilon bc} \tag{1.1}$$

where I_0 and I_t are the intensities of the incident and transmitted light beams, ϵ is the molar absorptivity (1 mol^{-1} cm^{-1}), b is the path length in cm, and c the molar concentration. If it is assumed that the intensity of the absorbed light I_a is given by $I_0 - I_t$ then

$$I_a = I_0 (1 - 10^{-\epsilon bc}) \tag{1.2}$$

The intensity of the fluorescence I_f is given by $I_a \phi_f$ (see Section 1.1 above), so

$$I_f = I_0 \phi_f (1 - 10^{-\epsilon bc}) \tag{1.3}$$

$$= I_0 \phi_f \left\{ 1 - [1 - 2.303 \, \epsilon bc + \frac{(2.303 \, \epsilon bc)^2}{2!} - \ldots] \right\} \tag{1.4}$$

If $\epsilon bc \leq$ about 0.05 all the terms in the expansion of $10^{-\epsilon bc}$ after $-2.303 \, \epsilon bc$ can be ignored, yielding

$$I_f = 2.303 \, I_0 \phi_f \, \epsilon bc \tag{1.5}$$

This oft-quoted expression is useful in demonstrating that (a) $I_f \propto I_0$ i.e. intense light sources should yield intense fluorescence (but see Chapter 6), (b) the fluorescence from a compound depends on its ϵ value as well as its quantum yield and (c) that $I_f \propto c$ if all the assumptions used in deriving equation (1.5) are valid. In practice, closer consideration of equation (1.4) shows that substantial errors can occur even in very dilute solutions. As a rough guide a solution whose absorbance A (= ϵbc) is 0.01 will generate approximately a 1% error in I_f compared with equation (1.5); a solution with $A = 0.05$ (e.g. a 5 μM solution of a sample with $\epsilon = 10^4$ in a 1 cm cell) will produce an error of about 5%, and so on [5]. It cannot be too strongly emphasized that such deviations from linear analytical curves are caused solely by the breakdown of the mathematical assumptions, and are in no way connected with the 'inner filter' effects (*optical artefacts*) described in Chapter 5. In other words, analytical curves may be non-linear for two entirely separate reasons. It seems likely that, in practice, analytical curves are non-linear in many cases: the advent of microcomputers that can readily be interfaced with optical instruments will presumably facilitate the use of 'curve corrections' analogous to those used in atomic absorption spectrometry and other methods. A test method for the linearity of fluorescence systems has been described [6].

1.6 Nomenclature in fluorescence spectrometry

In any technique standardization of working procedures implies standardization of nomenclature; alas, the nomenclature in this field is confused as it is in many others. In particular the terms *fluorimeter* and *fluorimetry* are in common use in the United Kingdom, whereas *fluorometer* and *fluorometry* are normal in the USA. We have followed the recommendation of Mielenz [7] in describing an instrument for measuring fluorescence spectra and intensities as a *fluorescence spectrometer*. In view of the almost universal use of the term *quantum yield* we have retained this expression rather than use the recommended alternatives *radiant yield* or *photon yield*.

Other spectroscopic nomenclature follows, wherever possible, the recommendations of the American Society for Testing and Materials [8].

References

1. Burgess, C. and Knowles, A. (Eds.) (1981), *Standards in Absorption Spectrometry,* Chapman and Hall, London.
2. Wheeler, J. P. (1977), *Ind. Research,* 7.
3. Shute, D. J. (1980), *Med. Lab. Sci.,* **37**, 173.
4. Leaback, D. H. (1976) *An Introduction to the Fluorimetric Estimation of Enzyme Activities,* 2nd edn, Koch-Light Laboratories Ltd, Colnbrook.
5. Parker, C. A. (1968), *Photoluminescence of Solutions,* Elsevier, Amsterdam.
6. *Linearity of Fluorescence Measuring Systems* (1980), American Society for Testing and Materials, publication no. E578-76.
7. Mielenz, K. D. (1976) *Analyt. Chem.,* **48**, 1093.
8. *Standard Definitions of Terms and Symbols Relating to Molecular Spectroscopy* (1980), American Society for Testing and Materials, publication no. E131-80.

2 Monochromator wavelength calibration

2.1 Introduction

The essential components of a single beam fluorescence spectrometer without facilities for spectral corrections are shown in Fig. 2.1. The instrument has two monochromators or alternative wavelength-selection devices. The excitation monochromator (M1) selects the

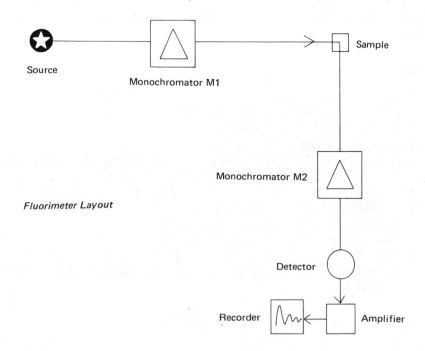

Fig. 2.1 Basic layout of a fluorescence spectrometer providing uncorrected spectra.

wavelength of the light with which the sample is to be irradiated, and the emission monochromator (M2) selects the wavelength of fluorescence (or scattered light or phosphorescence, etc.) received by the detector. M1 and M2 may be coloured or interference filters in simple instruments, but in more advanced instruments grating monochromators are normal. Some instruments incorporate double-grating monochromators to minimize stray light levels (see Chapter 3). The monochromators may normally be scanned using motorized drive units. Scanning M1 with M2 set at a fixed wavelength generates excitation spectra and scanning M2 with M1 fixed generates fluorescence emission spectra. In view of the increasing interest in synchronous scanning methods [1] the facility to scan both monochromators simultaneously is desirable (see also Section 7.1.3). The reflection gratings used in M1 and M2 are normally blazed to give maximum efficiencies (approximately 50%) at 250–350 nm and 350–500 nm, respectively. Holographic gratings, which yield low stray-light levels (see Chapter 3) have been introduced in some instruments. Several instruments use continuously-variable, narrow bandwidth interference wedges as simple monochromators. These may also be motor-driven to provide (normally emission) spectra.

Fluorescence spectrometers containing grating monochromators may also incorporate filters to eliminate second order effects. For example, if the emission monochromator is set at 500 nm, it will also pass Rayleigh-scattered exciting light of 250 nm wavelength. This signal can be removed by provision of a suitable cut-off filter.

It is clearly desirable that, for all fluorescence measurements, the monochromators are correctly adjusted. Accurate adjustment is obviously essential if corrected spectra are to be obtained and, in quantitative studies, sensitivity and selectivity may be sacrificed if the monochromators are not at the optimum wavelength setting. For most purposes an accuracy of ±1 nm throughout the range of the monochromator is sufficient. In practice, although the specifications of commercially-available instruments sometimes claim accuracies of 1–2 nm or better, errors of 10 nm or more can arise.

2.2 Characteristics of calibration methods

A suitable wavelength calibration technique should have the following characteristics:

(a) Since monochromators may go out of adjustment relatively frequently, they should be calibrated regularly, and a quick and

10 Standards in fluorescence spectrometry

convenient method is thus desirable.

(b) The calibration procedure should be available over a range of wavelengths: the corrections to be applied are not necessarily equal over the entire spectral region covered by the monochromator.

(c) The procedure should be applicable in the conditions in which the instrument is generally to be used e.g. with the light source in the usual position and with realistic spectral bandwidths.

It should be noted that it is necessary only to calibrate one monochromator initially. The other monochromator can then be calibrated using the Rayleigh scattered light from a sample in the normal sample position. Colloidal silica solutions have been recommended for this purpose [2], but the signal from distilled water may be quite adequate (see Section 4.3). It is important to note that the accuracy of the excitation monochromator can be affected by the position of the arc in the light source [2]. In addition, if the calibrations are performed with the aid of a pen-recorder, the recorder itself can introduce errors unless it is properly calibrated and the chart paper moves at a constant speed [3]. Backlash on the manual monochromator controls can also be a source of inaccuracy.

In practice, three approaches to the problem of wavelength calibration have been used, namely:

(a) Use of sharp spectral lines in the output of the instrument's own light source or lines obtained with the aid of filters.

(b) Use of sharp spectral lines from an auxiliary light source.

(c) Use of sharp fluorescence peaks in the spectra of organic or inorganic species, either in solution or in synthetic polymeric glasses.

These methods will now be considered in turn.

2.3   Use of spectral lines from the spectrometer light source

Fig. 2.2 shows the output of a 150 W xenon lamp typically used in fluorescence spectrometers. The sharp lines occur in the region 400-500 nm: amongst the most prominent are those at 450.1, 462.4, 467.1 and 473.4 nm [4]. (More intense lines occur at >800 nm but these are of no value in conventional studies.) The pulsed xenon sources fitted in several recent instruments have similar emission spectra [5]. These lines can be used to calibrate the excitation monochromator with the aid of either a quantum counter (cf. Section 7.2) or by using a reflector plate with the emission monochromator set for the zero order. Recent work [6] shows that

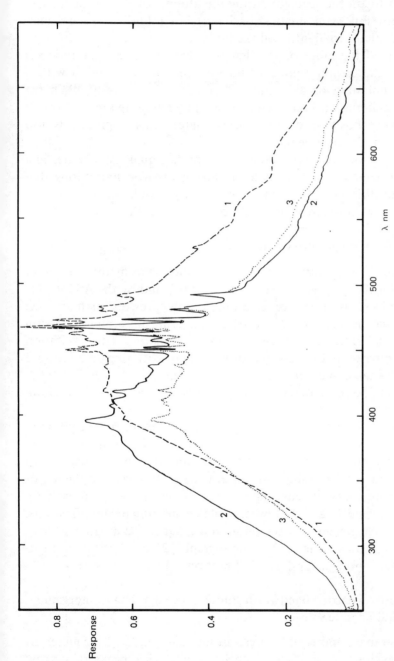

Fig. 2.2 Relative spectral power outputs of three 150 W xenon lamps Curve 1: lamp supplied by manufacturer A, studied after only a few hours use; curve 2: lamp supplied by manufacturer B, new condition; curve 3: lamp of same type as curve 2, after some hours use. (Data kindly supplied by R. Burns, Division of Mechanical and Optical Metrology, National Physical Laboratory).

the positions and relative intensities of these lines may vary (a) from one lamp to another and (b) during the lifetime of a single lamp. The wavelength shifts are small (~ 0.5 nm) but intensity changes may make it difficult to identify all the lines.

To extend the range of wavelengths over which the calibration is available, the xenon source could be used in conjunction with a standard holmium or didymium filter [7]. Alternatively, discharge lamps containing different elements could be mounted in the normal light source position of the spectrometer: this approach is too inconvenient for routine use.

A number of instruments that incorporate a quantum counter in a reference channel also include arrangements for monitoring the output of this channel alone: this facility would be of value in connection with the methods described in this section.

2.4 Use of an auxiliary light source

A second procedure for aligning monochromators is that recommended by Chen [2], by Guilbault [8] and by the ASTM [9]. This involves inserting, for example, a low pressure mercury pen lamp in the normal sample position and using its atomic emission lines to calibrate the emission monochromator. The excitation monochromator is then aligned using a light-scattering solution (see above). Suitable mercury lines are those at 253.7, 296.7, 302.2, 313.2, 334.2, 366.3, 404.7, 407.8, 435.8, 546.1, 577.0 and 597.1 nm [8].

This method allows the emission monochromator to be calibrated over a wide wavelength range, but it has the disadvantage that in practice the emitted intensity of the Hg lamp is much too high. Even with the aid of a neutral density filter (Chen [2] suggested the use of several thicknesses of lens tissue!) it may be necessary to use very narrow bandwidths and the lowest available instrument amplifications. Without such precautions the photomultiplier in the instrument may become saturated. The ASTM monograph [9] emphasizes that slit-width effects may occur during calibration.

2.5 Use of narrow bandwidth fluorescence maxima of inorganic and organic solutes

A considerable number of fluorescent inorganic and organic materials exhibit excitation and fluorescence spectra with peaks of narrow bandwidth. These sharp maxima can be used to calibrate the monochromator of a fluorescence spectrometer; spectral correction effects are negligible in such cases.

Amongst inorganic species, the trivalent lanthanide ions provide suitable spectra. Terbium (III) and europium (III) in particular exhibit readily-detectable luminescence. The bands in the Tb(III) spectrum occur as a result of transitions from the $^5D_4$ excited state to various components of the ^7F ground level. The principal emission wavelengths are at 490 and 545 nm: excitation can take place at any of a number of maxima in the excitation spectrum between 260 and 400 nm. The 490 nm and 545 nm bands have spectral bandwidths not exceeding about 15 nm. The intensity of the terbium luminescence can be greatly enhanced by numerous chelating agents, nucleic acids and proteins (e.g. [10, 11]) but the enhancement effects may sometimes be accompanied by small wavelength shifts [11], so this procedure cannot be recommended. In the cases of europium (III) solutions, the principal emission bands are centred at 585 and 614 nm, with the optimum excitation wavelength 394 nm. Other lanthanides are also luminescent — for example Gd(III) has a single intense emission band at 312 nm — but it should be noted that the fluorescence of the cerium (III) ion is a broad band emission unsuitable for calibration purposes.

Glassy solids doped with lanthanide ions make excellent standards for wavelength calibration if they can be fashioned into blocks of standard cuvette shape. The properties of these glasses have been studied in detail by Reisfeld [12] who has shown that the emission wavelengths are dependent to a small degree on other (inorganic) components of the glass. Thus the thulium (III) lines that occur at 705, 456 and 355 nm in borate glasses occur at 701, 453 and 350 nm in phosphate glasses; and in borate glasses the principal Tb(III) bands are at 486 and 541 nm, about 4 nm lower than in aqueous solutions of the ion.

Amongst organic compounds, aromatic hydrocarbons seem to be the most suitable for wavelength calibrations. Many show an intense fluorescence with well-marked vibrational fine structure. They can be used in solution, but are also commercially available in acrylic copolymer matrices. These matrices, supplied with representative spectra and machined to fit standard cuvette holders, are very stable. Moreover, in such an environment, certain hydrocarbons and other molecules exhibit room temperature phosphorescence as well as fluorescence and can thus be used over a wide range of wavelengths. (To prevent oxygen quenching of the triplet state, such standards should be stored under nitrogen. Oxygen can be removed by heating to 100° C *in vacuo*. Glasses which fit Dewar flasks, and can thus be used at 77 K, are available: in these conditions, oxygen quenching is not a serious problem). Coronene is a typical example, with well-defined fluorescence maxima at 426, 434, 446, 453 and 474 nm, and phosphorescence maxima at 499, 511, 529, 552 and 566 nm.

Other possible fluorescent wavelength standards include ovalene and anthracene-naphthalene mixtures [7]. Solutions of inorganic or organic species in sealed cells make excellent alternatives to the glassy solids and are less costly.

2.6 Conclusions and recommendations

From the preceding paragraphs it is apparent that those inorganic or organic fluorescent materials which have narrow emission bands or well-defined vibrational fine structure provide the most convenient and acceptable method for routinely checking the monochromator settings of a fluorescence spectrometer. They can be selected to cover a wide range of wavelengths and their use requires neither modification of the instrument nor the introduction of extra equipment. Effective calibration can also be obtained by using the instrument's own light source, possibly with the aid of additional filters. The previously-recommended method [2, 8, 9] involving the use of an auxiliary light source appears to be the least desirable approach.

References

1. Lloyd, J. B. F. (1971), *Nature Phys. Sci.*, **231**, 64.
2. Chen, R. F. (1967), *Anal. Biochem.*, **20**, 339.
3. Hercules, D. M. (1957), *Science*, **125**, 1242.
4. *Handbook of Chemistry and Physics* 55/e (1974), R. C. Weast, ed., CRC Press, Cleveland, Page E-210.
5. West, M. A. (1975), *Int. Lab.*, Jan/Feb, 41.
6. Burns, R. (1979), private communication.
7. West, M. A. and Kemp, D. R. (1976), *Int. Lab.*, May/June, 27.
8. Guilbault, G. G. (1973), *Practical Fluorescence*, Marcel Dekker, New York.
9. Annual Book of ASTM Standards (1972), Part 42 Standard E388-72, American Society for Testing and Materials.
10. Kayne, M. S. and Cohn, M. (1974), *Biochemistry*, **13**, 4159.
11. Luk, C. K. (1971), *Biochemistry*, **10**, 2838.
12. Reisfeld, R. (1973), *NBS Special Publication* no. 378, 203.

3 Stray light in fluorescence spectrometers

3.1 Origins of stray light and resultant errors

Stray light may be defined [1] as light of unwanted wavelengths emerging from a grating monochromator or other dispersion device: light of the selected band-width is thus contaminated by the stray light. In general, stray radiation can arise from a variety of causes [2]:

(a) Leakage of external light into the fluorescence spectrometer.
(b) Reflection and scattering from walls, optical surfaces, and other instrument components.
(c) Scattering of light by airborne dust.
(d) Scattering within optical components such as lenses.
(e) Background fluorescence of optical materials.
(f) Unused orders in grating spectra.

Many of these effects can be minimized by careful use and maintenance of the instrument. Stray light from grating monochromators is considered below.

Stray radiation is clearly undesirable in all forms of fluorescence spectrometry. In particular, in cases where very low luminescence intensities must be measured, the stray light may make an important contribution to the background signal and hence to the limits of detection available. The problem will be most severe when excitation wavelengths in the ultra-violet region (where conventional light sources are feeble) are used. The intensity of stray light in the visible region may then be considerable, although it can be reduced with the aid of suitable filters. Chen [1] has pointed out that stray light could be a source of a similar error during calibration of a spectrometer light source using the Rhodamine B screen method (see Section 7.2.2). Since this compound absorbs light over a wide wavelength range, stray visible light could lead to misleading results while the calibration curve in the ultra-violet region is being constructed. Such errors can be checked by using an alternative material such as dimethylaminonaphthalene sulphonate as a fluorescent screen: this compound has little absorption at wavelengths above 400 nm. A further possible error due to stray radiation has been highlighted by Parker [3] who has pointed out that anthracene absorbs the 254 nm mercury line over 100 times more strongly than it absorbs the 313 nm line of the same element. Thus, if the 313 nm line is isolated by a filter or monochromator that transmits 1 per cent of the intensity of the 254 nm line, the observed fluorescence intensity will be twice as great as would be observed in the absence of the stray 254 nm radiation. Such effects could cause large errors in, for example, the determination of quantum yields using the comparison method (see Section 8.3).

Despite the evident importance of stray light in luminescence spectroscopy, many instrument specifications entirely ignore this factor. In other cases, stray light levels are only given at one or two wavelengths. The methods by which stray radiation is determined are rarely cited. Stray-light levels, measured as a proportion of total light throughput, are generally highest at the lowest wavelengths and some manufacturers are thus content to give a stray-light specification at a low wavelength, e.g. 230 nm.

3.2 Stray light in grating monochromators

Commercially-available instruments almost invariably utilize grating monochromators. They have many advantages over prism monochromators, and only one substantial disadvantage, that of multiple orders (see above). Stray radiation arises in such monochromators as a result of flaws and irregularities on the grating surfaces: a detailed study has been published by Sharpe and Irish [4]. Until recently, plane reflection gratings, blazed for maximum efficiency at suitable wavelengths, have been generally used, often mounted in the Ebert or Czerny-Turner configurations (Fig. 3.1a) which conveniently minimize various optical aberrations. The stray light from a small monochromator of this type may be as low as 0.1–1 per cent of the total light flux at the exit slit. Much lower stray-light levels can be obtained by using *double grating* monochromators. These consist essentially of two monochromators mounted in series, the exit slit of the first being the entrance slit of the second (Fig.

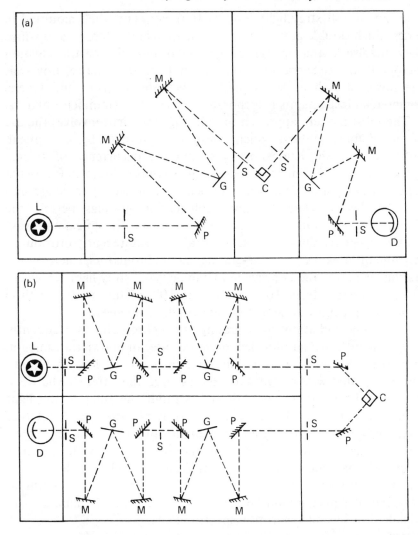

Fig. 3.1 *Simplified optical diagrams of fluorescence spectrometers containing (a) two single grating monochromators, and (b) two double-grating monochromators. The monochromators shown are of the Czerny-Turner type: an Ebert monochromator is similar, with a single large collimating mirror, replacing the pair of small concave mirrors. Optical components are lettered: C, sample cuvette; D, detector; G grating; L, light source; M, concave mirror; P, plane mirror; S, slit.*

3.1b). In order to keep the two components wavelength-synchronized, the two gratings are normally mounted on the same shaft or rotating table. Stray radiation in such cases will be drastically reduced; two gratings each giving stray-light levels of one per cent will ideally yield an overall stray-light level of 0.01 per cent when mounted in series. Such double-grating monochromators are expensive and (since they involve extra optical components) have lower transmission factors than single-grating monochromators. In practice, however, the lower stray-light levels may compensate for this, and several commercial instruments incorporate double grating monochromators.

The most modern approach to reducing stray light involves the use of holographic gratings. Such gratings are formed by an optical, rather than a mechanical process, and the occurrence of tracing defects is thereby eliminated. Stray light is thus reduced to 0.001 per cent or less for a single-grating monochromator. Holographic gratings can be blazed, and concave holographic gratings permit the elimination of a number of plane mirrors from the optical pathway, with a higher resultant transmission factor. One recently-introduced instrument uses two double-grating monochromators, each of the gratings being of the concave holographic type: stray light is claimed to be between one in 10^{10} and one in 10^{14} of the total light flux, though no precise wavelength-dependent data are given.

The stray radiation in a grating monochromator is frequently measured with the aid of a glass or liquid filter of known transmission properties [5]. When the monochromator is set to a low wavelength and illuminated with a continuum light source, the difference between the transmitted intensities with and without the filter in the light beam is a measure of the proportion of stray radiation present. Other filters may be used for higher wavelength work and a similar method can be adapted for determining the stray light in fluorescence spectrometers. Since stray radiation is strongly wavelength-dependent, the stray-light levels should be cited at two or more wavelengths, at least one of which should be below $\leqslant 250$ nm. Such data are already available for several absorption spectrometers.

3.3   Summary and recommendations

(a) The stray light properties of instruments should be carefully specified by their manufacturers.

(b) Such a specification should cover a range of wavelengths, and should include a brief description of how, and at what point in the light path, the stray-light levels are measured.

(c) The operator can minimize additional stray-light effects by careful maintenance of his instrument and careful experimental technique. Suitably placed filters may also be of value.

References

1 Chen, R. F. (1967), *Anal Biochem.*, **20**, 339.
2 Winefordner, J. D., Schulman, S. G. and O'Haver, T. C. (1972), *Luminescence Spectrometry in Analytical Chemistry*, Wiley-Interscience, New York.
3 Parker, C. A. (1968), *Photoluminescence of Solutions*, Elsevier, Amsterdam.
4 Sharpe, M. R. and Irish, D. (1978), *Optica Acta*, **25**, 861.
5 Annual Book of ASTM Standards (1972), Part 42 Standard E387-72. American Society for Testing and Materials.

4 Criteria for fluorescence spectrometer sensitivity

4.1 Background: inter-instrument comparisons

All manufacturers of filter and grating fluorescence spectrometers include in each instrument specification some estimate of the sensitivity of the instrument and/or the limits of detection obtainable by its use. Some typical examples are given in Table 4.1. Such specifications are, in principle, desirable for two reasons:

(a) Most workers using fluorescence methods are attracted principally by their sensitivity.

(b) To a far greater extent than in (for example) UV–visible absorption spectrometry, the sensitivity of a fluorescence assay is dependent on the instrument in use (and on its condition and many other factors - see below).

Some means of comparing instrument performance, and of checking that a particular instrument is performing satisfactorily, are thus clearly necessary. The considerable difficulties involved in satisfying this need are outlined in the following paragraphs.

The concept of sensitivity in luminescence spectroscopy has been discussed in detail by Parker [1] and by Winefordner *et al* [2]. Parker [1] distinguished three aspects of sensitivity, namely:

(a) *Instrumental sensitivity*

This term refers to the performance of a particular instrument, and is the context in which we use the term *sensitivity* in subsequent paragraphs of the present discussion. It may be defined by (i) giving the signal-to-noise ratio of the instrument in a particular closely-defined set of operating conditions or (ii) expressing the minimum detectable quantity of a particular compound (usually quinine

Table 4.1: *Some examples of instrument sensitivities cited by manufacturers*

Instrument	Sensitivity cited	Comments (see text)
1	Minimum detectability 0.5 x 10^{-9} g. ml^{-1} quinine sulphate in 0.1 N H_2SO_4.	Minimum detectability not defined: experimental conditions not given: quinine sulphate not an ideal standard. See also 3.
2	Sensitivity: 40 ppt.	Valueless: no experimental information given, not even the identity of the sample!
3	Limit of detection is less than 0.5 parts per trillion of 9, 10-diphenylanthracene in cyclohexane, measured with 5 nm emission bandwidth and 20 nm excitation bandwidth. These limits are for peak to peak noise equal to solution minus blank.	More experimental information (but not λ_{ex} and λ_f) given, and limit of detection is defined. Limited applicability in comparisons with other instruments, or in studies of other solutes. L.o.d. is affected by background signal from cell, solvent etc.
4	Raman band of water: λ_{ex} = 350 nm, λ_{fl} = 397 nm; slits 10 nm both monochromators; S/N not less than 40:1.	Superior approach in theory: not very meaningful in practice: only of value at the wavelengths cited.

sulphate in practice), also in standard conditions (see (c) below). Procedure (i) is much to be preferred, for reasons given below.

(b) *Absolute sensitivity*

This term relates to the sensitivity of a particular material and is expressed entirely in terms of the intrinsic properties of that material, namely, its extinction coefficient, quantum yield and spectral bandwidth. Again, the experimental conditions, for example solvent, temperature, have to be carefully defined. Multiplication of the absolute sensitivity by an instrumental factor generates an instrument signal value [2].

(c) *Method sensitivity*

This is more usually referred to as the limit of detection and naturally incorporates conditions from both the intrinsic properties of the luminescent sample, and from the instrument performance. In practice it is also affected to a great extent by other factors (see Section 4.2). It is most important to appreciate that the *limit of detection* of a particular procedure is itself a matter of definition. In trace analysis, the limit of detection is frequently defined as that concentration that yields an instrument response two standard deviations above the background signal (see, for example, [3]). Several other definitions, including some based on the t-statistic, have been used [2, 4].

4.2 The limit of detection method

The limit of detection of a particular solute is dependent in practice not only on the intrinsic luminescence properties of that solute, and on the performance of the fluorescence spectrometer at the appropriate excitation and emission wavelengths, but also on several other factors. These include Raman scattering from the solvent, the luminescence of the cuvette (see Section 5.2.2), luminescence from impurities in the solvent and luminescence from impurities in the sample. The interference from the Raman effect can sometimes be minimized by the choice of a suitable solvent, and the luminescence of the cuvette and of the solvent can be minimized by the use of the highest quality materials and by meticulous cleanliness in experimental work. Nonetheless, in studies of a single component of a complex mixture (the usual type of fluorescence experiment) the limit of detection is likely to be *blank-limited* [1], that is, set by the luminescence and scattering signals from sample components other than the solute. In practice, therefore, if two instruments are compared, their instrumental sensitivities and the limits of detection attainable may be quite irreconcilable. Thus Parker and Barnes [5] compared a filter fluorescence spectrometer and a monochromator instrument. The instrumental sensitivity of the former (expressed as the equivalent concentration of quinine sulphate) was *calculated* to be 100 times better than that of the latter. The limit of detection actually attainable for quinine sulphate was, however, some 4 times better (lower) on the grating instrument than on the filter instrument, since the filter spectrometer gave a high blank reading.

Had it been necessary to determine the quinine sulphate in a complex sample matrix, the advantage of the monochromator instrument might well have been even greater. It must be concluded that expressing the sensitivity of an instrument in terms of a hypothetical limit of detection for a single solute may give such a misleading impression as to be valueless. Furthermore, this practice only provides (dubious) information on instrument performance at a particular pair of excitation and emission wavelengths: two instruments whose performances at the appropriate wavelengths for quinine sulphate, for example, are similar may not exhibit such a similarity when another solute is studied at different wavelengths.

4.3 The signal-to-noise ratio method

The alternative approach to citing instrument sensitivity, that of providing the signal-to-noise ratio of the spectrometer in a well-defined set of conditions, has recently become more popular amongst instrument manufacturers, and is certainly preferable to the *limit of detection* approach. Parker [1] was apparently the first to suggest that the Raman signal from a suitable solvent provides a useful test of instrument sensitivity (and, if necessary, resolution). The procedure involves determining the signal-to-noise ratio of the instrument at particular slit-width and amplifier settings, with the excitation and emission monochromators set to appropriate wavelengths to detect Raman-scattered light from the solvent. In practice, water is usually used as the solvent: it can be obtained virtually free of fluorescent impurities by repeated distillation from silica apparatus and ultrafiltration. The principal Raman band of water occurs at 3380 cm^{-1} [6].

The excitation and emission wavelengths usually employed at present are 350 and 397 nm, with spectral bandwidths of 5 or 10 nm (see Table 4.2 for examples). In these conditions a sensitive instrument in good condition will normally have a signal-to-noise ratio of at least 30:1.

Table 4.2: *Instrument sensitivities expressed as signal-to-noise ratios*

Instrument	λ_{ex} (nm)	λ_{fl} (nm)	S_{ex} (nm)	S_{fl} (nm)	t (s)	$d\lambda/dt$ (nm s^{-1})	S/N
1*	350	397	10	10	2	—	$\geqslant$40:1
2	350	397	10	10	1	—	$\geqslant$30:1
3	350	397	5	5	1.2	1	$\geqslant$35:1

*Same instrument as number 4 in Table 4.1. S_{ex} S_{fl} are the excitation and fluorescence monochromator bandwidths, t the instrument time constant, and $d\lambda/dt$ the scan rate.

Cyclohexane has also been recommended as a suitable material for the detection of Raman-scattered light, though the ASTM procedure [7] suggests that the minimum detectable concentration of this material in chloroform should be determined from the Raman signal.

Although these procedures avoid some of the problems of the *limit of detection* approach, they are not free from drawbacks. In particular it is not sufficient to perform the test at a single pair of excitation and emission wavelengths. Different spectrometers differ in their relative responses at different wavelengths (see above). In addition a single instrument will show changes in sensitivity with time. Such changes, which will clearly occur at a rate dependent on the conditions in which the instrument is housed and the care with which it is used and maintained, most frequently involve a more rapid deterioration in performance at wavelengths below 300 nm than at higher wavelengths. (This deterioration is due to the gradual coating of the light source envelope and other optical surfaces with UV-absorbing contaminants).

Determining the signal-to-noise ratio at only a single pair of excitation and emission wavelengths will thus give incomplete information, both in comparisons of two or more instruments and in monitoring the performance of a single instrument.

The signal-to-noise ratio should be determined at several pairs of excitation and emission wavelengths. The excitation wavelengths should correspond to bright lines in the mercury spectrum: the method could then be applied to simple instruments fitted with mercury lamps as well as to more advanced spectrometers fitted with xenon lamps. Suitable excitation wavelengths would be 254, 313, 366 and 436 nm; the corresponding Raman wavelengths for water would be 278, 350, 418 and 511 nm; it is essential that the spectral bandwidth, scan rate, time constant and other relevant instrument parameters (for example, the insertion of supplementary filters) should be provided, along with the signal-to-noise values.

If such data are to be of any value in comparing the performance of one instrument with another, it is clearly of importance that these operating conditions should be as similar as possible. Unfortunately, instrument design differences may prevent this. For example, the specification for a spectrometer in the author's laboratory claims a signal-to-noise ratio of at least 35:1 with a scan rate of 1 nm s^{-1}, excitation and emission bandwidths of 5 nm, and a time constant (response time) of 1.2 s. Another instrument in the same laboratory can be operated at a scan speed of 1 nm s^{-1}: but none of the available (fixed) slits provide bandwidths of 5 nm and a time constant of 1.2 s is not available. A direct and valid comparison between these instruments, using the signal-to-noise ratio criterion, is thus not possible.

A final disadvantage of the signal-to-noise ratio approach, perhaps more apparent than real, is its lack of immediate relevance to everyday fluorimetry.

A statement that an instrument has a particular signal-to-noise ratio in a particular set of operating conditons may mean little to many users of the technique, in terms of their own studies. A number of examples in the literature further suggest that many workers are totally unaware of the likely presence of Raman signals in the spectra generated in fluorescence studies. In these circumstances, there must be a danger that this approach to expressing instrument sensitivities will be of limited practical value, in spite of its intrinsic advantages.

4.4 Summary and recommendations

(a) There seems to be no ideal and readily-available method which can be used to express the sensitivity of an individual fluorescence spectrometer or to compare the performances of different instruments.

(b) The most suitable method of expressing sensitivity is the signal-to-noise ratio method, using the water Raman band. A series of appropriate excitation and emission wavelengths, the excitation wavelengths corresponding to the bright lines in the mercury emission spectrum, should be used, and the minimum signal-to-noise ratio given in each case. Additional essential information includes the spectral bandwidth, time constant and scan rate used during the test.

(c) The limitations of this procedure must be constantly borne in mind. In particular, the time-dependent changes of spectrometer performance must be remembered: the sensitivity of an instrument may in practice be substantially less than any value claimed by the manufacturer.

References

1. Parker, C. A. (1968), *Photoluminescence of Solutions*, Elsevier, Amsterdam.
2. Winefordner, J. D., Schulman S. G. and O'Haver, T. C. (1972), *Luminescence Spectrometry in Analytical Chemistry*, Wiley-Interscience, New York.
3. Pecsok, R. L., Shields, L. D., Cairns, T. and McWilliam, I. G. (1976) *Modern Methods of Chemical Analysis*, 2nd edn., John Wiley, New York.

4 Curry, S. H. and Whelpton, R. (1978), in *Blood Drugs and Other Analytical Challenges* (E. Reid, ed.,), Ellis Horwood, Chichester.
5 Parker, C. A. and Barnes, W. J. (1957), *Analyst* (London), **82**, 606.
6 Parker, C. A. (1959), *Analyst* (London) **84**, 446.
7 Annual Book of ASTM Standards (1976) Part 42 Standard E579-76, American Society for Testing and Materials.

5 Inner filter effects, sample cells and their geometry in fluorescence spectrometry

5.1 Inner filter effects

5.1.1 *Introduction: consequences of inner filter effects*

Under commonly used experimental conditions, the observed fluorescence signal from a solution, relative to the concentration of the fluorophore, generally decreases as the solution becomes increasingly concentrated. The decrease is due, in part, to an attenuation of the excitation beam in regions of the solution in front of the point from which the fluorescence is collected by the detection system, and to absorption of the emitted fluorescence within the solution. The effect is the *inner filter effect* as defined, therefore, by Parker and Rees [1] and includes all light-attenuating processes - due not only to the fluorophore but also to any other chromophores that may be present. Other terms that have been used in this context include *self-absorption*, *screening*, *prefilter* and *past filter effects*, and the *trivial effect*. The latter term belies the effect's importance both in nature, and in the laboratory [2].

Although the phenomenon has been known in various forms for many years [3], and although most textbooks dealing with fluorescence spectroscopy draw attention to it, results and spectra continue to be published that are subject to the effect, sometimes to a massive and apparently unsuspected degree. This could be avoided, or at least recognized, if the recommendations of Chapman *et al.* [4] for the standardized reporting of spectra were always adopted. No standardization in the subject is possible if inner filter effects are not recognized and counteracted as far as possible.

There are, of course, other concentration-dependent processes that reduce fluorescence intensities: intermolecular processes involving either or both ground and excited states of a fluorophore are a common cause. Singlet state-quenching molecules may be present adventitiously, for instance oxygen in aerated solvents, or heavy atom perturbers such as residues of halogenated solvents. Even so, it is apparent that the most common source of error and misinterpretation is the inner filter effect; and Parker's view that 'lack of appreciation of the elementary principles governing the inner filter effect ... was one of the main reasons why the technique of fluorescence measurement has been regarded with suspicion by some workers' [5] unfortunately is still relevant today.

The major consequences of inner filter effects are as follows:

(a) As the concentration of a fluorophore is increased, the initial, linear growth of fluorescence with concentration will fall away; with some types of cell geometry, the fluorescence signal will decrease. Typical plots are given by Udenfriend [6] and by Guilbault [7]. An increasing concentration of non-fluorescent chromophore similarly suppresses the fluorescence signal from a constant concentration of fluorophore. The fluorescence emitted by the final solutions increases when they are diluted, provided other experimental variables are held constant.

(b) Observed fluorescence intensity decreases as the point from which it is monitored is moved further away from the illuminated surface of the solution. The effect has been used to distinguish inner filter from quenching effects, for example, by Brandt *et al.* [8]. In concentrated solutions, fluorescence emission is restricted entirely to the vicinity of the illuminated surface.

(c) Where several bands differing in amplitude are present in the absorption spectrum of a compound, those having relatively high absorption coefficients will be depressed relative to the remainder in the fluorescence excitation spectrum unless adequately dilute solutions are used [1].

(d) Similarly, fluorescence emission spectra excited in regions of high absorbance will be suppressed. The effect is seen in mixtures [9] as well as in single compounds. Impure, light-absorbing solvents also give rise to the effect.

(e) A fluorescence emission spectrum will be weakened at wavelengths where overlap with the corresponding absorption (excitation) spectrum occurs, and observed fluorescence lifetimes are increased [10]. An emission spectrum is similarly weakened by any other coincident absorption which, if structured, will appear as inverted peaks in the spectrum [1]. To casual examination, the result may seem to be a genuinely structured fluorescence spectrum.

(f) Similar interferences may arise in excitation spectra when coincident extraneous absorption is present.

(g) Absorbed fluorescence may be re-emitted as *secondary fluorescence*, which results in anomalous quantum yields [11].

5.1.2 Cell geometry and inner filter effects

The magnitude and characteristics of inner filter effects depend strongly on sample-cell geometry, as discussed by Parker [5]. Usual geometries, and their consequences, are:

(a) *Front surface geometry*

In front surface geometry the fluorescence is collected and measured through the illuminated surface. The arrangement is generally used when the emission spectrum of a strongly absorbing solution is required. Under these conditions virtually all of the excitation is absorbed and the cell functions as a quantum counter. If excitation and emission spectra overlap, secondary emission occurs, for which the results must be corrected [12]. For dilute solutions, any advantages that the arrangement may have are largely discounted by the high level of scattered light diverted to the detector [13]; and if compared samples differ in absorbance then differences in the penetration of excitation into the cell contents will result in different emitting volumes being seen by the detector [14].

(b) *In line geometry*

In this geometry the fluorescence collected passes through the surface opposed to the irradiated one. The fluorescence becomes increasingly localized at the irradiated surface as the concentration of the solution increases and traverses practically the whole depth of the solution. It is, therefore, particularly subject to distortion at wavelengths of coincident absorption. (The large amount of exciting light entering the emission monochromator in this geometry can be eliminated by Cassegrain optics [15]). When such absorption is absent, the proportionality between fluorescence and concentration is reported to be direct over almost the whole range of light absorption [16]. The geometry has been used with a vertical optical axis, which enables the optical depth and, hence, the fluorescence emitted and collected to be varied simply by the variation of the volume of the solution in the cell. Under these conditions, the fluorescence signal depends linearly on the volume of the solution used over a tenfold range. [17]; but the usefulness of such arrangements is greatly limited by their sensitivity to the inner filter effect of absorption of fluorescence and its re-emission. At high concentrations, the excitation spectra become that of a quantum counter.

(c) *Perpendicular geometry*

In perpendicular geometry the fluorescence emitted is collected along an axis at right angles to the excitation beam. Because of its freedom from the effects of large amounts of scattered and transmitted excitation, where inner filter effects can be countered by the use of dilute solutions, this is the preferred arrangement. Inner filter effects in concentrated solutions are severe in emission as well as in excitation spectra, in both of which the characteristic decrease in signal strength at increased concentrations occurs. As with in line geometry, the use of a vertical excitation axis enables the pathlength of excitation to be varied by adjustment of the volume of the sample [18, 19]. Other variable pathlength arrangements for horizontal excitation axes have been described [20, 21]. (From the variation of a fluorescence signal with pathlength, quantum yields may be determined [19, 21]).

5.1.3 *Countermeasures*

Optimally, all fluorescence measurements and spectra would refer to solutions at infinite dilution, where, given transparent solvents, absorption effects could be ignored. Although the great sensitivity of modern instrumentation often allows this limit to be approached sufficiently closely for many purposes, circumstances (apart from oversight!) sometimes occur where the use of relatively concentrated solutions is unavoidable. For instance:

(a) When the quantum yield of a fluorophore is very low.

(b) Although the quantum yield and concentration of a fluorophore are high, high levels of extraneous absorption coincident with the excitation spectrum are present.

(c) In an analysis, the use of a strongly absorbing reagent is necessary.

(d) When a state of equilibrium is undesirably affected by dilution.

(e) Where true quenching effects must be separated from the inner filter effect of a light-absorbing quencher.

On such occasions, the results may be corrected by factors calculated from the cell geometry and the absorption characteristics of the solution and/or experimental conditions may be chosen to counteract the effect as far as possible.

5.1.4 Correction factors

For discussions of the full relationships between the observed fluorescence signal and the optical and photophysical characteristics of a sample, reference is made to reviews by Melhuish [12], Lipsett [22], and Demas and Crosby [23]. Calculated corrections for inner filter effects are usually based on a considerable number of simplifying assumptions.

(a) Perpendicular geometry

A correction factor (C) is defined by which an observed fluorescence signal (F) at a specified wavelength can be corrected to the fluorescence signal (F_0) expected in the absence of the inner filter effect: $F_0 = CF$. By definition, F_0 is a linear function of concentration of the fluorophore.

In the simplest possible case, where it is assumed that the fluorescence is collected for measurement from a fixed point within the sample cell, then the excitation intensity at that point (I) is $I = I_0 \, 10^{-Ad}$ from the Beer-Lambert law, where I_0 is the intensity of excitation at the sample surface, A is the absorbance per cm pathlength at the relevant wavelength, and d is the depth in cm of the nominal pathlength of the excitation within the cell up to the monitored point. If the emitted fluorescence (F) is taken to vary linearly with I, and F_0 is similarly related to I_0, then $F = F_0 10^{-Ad}$ and $C \; (= F_0/F) = 10^{Ad}$. If the emission, indicated by the primed parameters, is also absorbed then:

$$C = 10^{(Ad + A'd')} \tag{5.1}$$

Thus, for a solution in a 1 cm cell with $A = 0.1$ and $A' = 0.05 \, \text{cm}^{-1}$ and $d = d' = 0.5$ cm, then $C = 10^{0.075} = 1.19$ at the wavelength concerned; and an uncorrected fluorescence would be subject to an error of 19 per cent.

The correction rests on a considerable number of assumptions and approximations: that the optical pathlengths can be defined; that the finite widths of the excitation and emission beams can be ignored; that the reflection of excitation and emission within the cell is insignificant; that the bandwidth of the light transmitted by the excitation and emission monochromators is infinitely small; that the optical characteristics of the instrumentation used for the absorbance measurements match those of the fluorescence spectrometer; that no change in refractive index occurs on dilution, and that no secondary fluorescence is present. No correction is made for light losses at the cell faces; the approximation implied is that they are constant over the wavelength range of any particular experiment.

Even so, for relatively dilute solutions, the relationship has apparently given satisfactory results. Thus, Weill and Calvin [24] report a linear relationship between the corrected fluorescence and concentration for solutions of proflavine with C-values of up to 1.52. Chen [25] used values for C varying up to 1.13 for solutions of quinine sulphate contained in short pathlength cells; his results seem to be free of inner filter effects.

The accuracy of the correction has been improved by Leese and Wehry [26], who make allowance for the variation in the absorbance of a sample over the wavelength range of the instrumental bandpass. In quenching studies of solutions with absorbancies varying up to about 1 cm^{-1} and with fluorescence pathlengths of 0.5 cm, good agreement of the derived Stern-Volmer coefficients with coefficients obtained from similarly corrected front surface geometry results is reported. Without the improved correction the coefficients decrease, but tend to the fully corrected values as the pathlengths are reduced. However, as only the quenching coefficients are reported, it is not possible to assess how well the correction performs over the different parts of the range of absorbancies used. The unmodified correction gives increasingly low results as the upper parts of this range are approached, under most conditions.

The equation has frequently been used to correct for absorption of excitation in a form modified to allow for the finite width of the excitation beam-delimited area viewed by the detector. Figure 5.1 shows the cell in cross-section.

The fluorescence seen, (F), is now proportional to the light absorbed between d_1 and d_2, and to the fraction of the total absorbance due to the fluorophore (y). Hence:

$$F = Ky(10^{-Ad_1} - 10^{-Ad_2})$$

where the constant K includes factors determined both by the photophysical constants of the sample and by the optical characteristics of the fluorescence spectrometer.

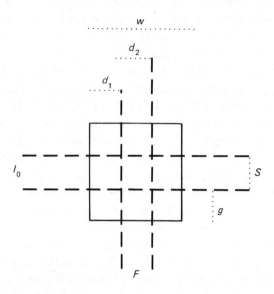

Fig. 5.1 *Cross-section of cell.*

As A becomes very small, F approaches F_0, and the terms containing d_1 and d_2 can be approximated by the first terms in their expansions.
Hence:

$$F_0 = Ky2.3A(d_2 - d_1)$$

And:

$$C = 2.3A(d_2 - d_1)/(10^{-Ad_1} - 10^{-Ad_2}) \qquad (5.2)$$

There is an appreciable number of examples of the use of this and similar equations [18, 19, 27-31]. Unfortunately, the accuracy of the correction is often not apparent; but Parker reports [32] that for values of C in excess of 3, the correction becomes increasingly inaccurate.

The factors limiting the application of the equation are discussed in detail by Parker and Barnes [27] and by Holland *et al.* [29]. Particular importance attaches to the requirements that the excitation must be a parallel beam, the cell geometry must be accurately defined, the fraction of the fluorescence detected and the angle at which it is detected must be constant over the whole of the viewed exit area of the cell, and the detector's sensitivity must be constant over all of its collecting surface. Using sample compartment optics designed to take account of these requirements, Holland et al. [29] corrected for the inner filter effect due to absorption of excitation in solution with absorbancies up to about 2 cm^{-1} to an accuracy of 3 per cent. Here, the corresponding values of C would vary up to approximately 6. Apart from the effect of the modified optics, however, it is significant that this improved accuracy is also partly due to the simultaneous determination of the fluorescence and absorbance spectra of the sample; and, apparently, by the empirical parameterization of the equation, for values of d_1 and d_2. These, however, were in good agreement with their nominal values.

An alternative expression for the dependence of fluorescence intensity on the absorption of excitation has been derived by Gill [33]. Inner filter effects in quinine sulphate solutions, having absorbances up to 1.36 cm^{-1}, could be corrected to an accuracy of 3 per cent. The relationship is used to show that for maximum sensitivity, which requires that the maxium absorption must occur behind the exit slit, an absorbance of about 1.0 is optimal in a 1 cm cell.

(b) *Empirical correction in perpendicular geometry*
Because inner filter effects are a function only of absorbance and cell geometry, and are independent of characteristics intrinsic to any particular sample, correction factors can be obtained entirely empirically, by least-squares curve-fitting procedures that linearly project the initially linear part of a fluorescence/concentration growth curve over regions where the curvature is severe. The approach has been used by Holland et al. [34], whose data acquisition system enables on-line corrections to be made. When sample absorbances are in the region of 2 cm^{-1}, the relative error is 7 per cent [34]. A graphical procedure for the correction of absorption of exciting light by an added quencher is described by Mertens and Kagi [35]: it requires that the relationship between log F and the concentration of an added non-fluorescent chromophore is linear when no quenching occurs.

(c) *Correction for absorption of emission in perpendicular geometry*
Because absorbances of common extraneous materials generally tend to decrease as wavelengths increase, inner filter effects are usually smaller in emission than in excitation spectra. The $10^{A'd'}$ factor has occasionally been used, but this makes no allowance for the finite

depth (S in Fig. 5.1) of solution from which the fluorescence is detected. An equation is given, without derivation, by Wu et al. [36]. But the equation requires that the correction factor should decrease and become negative as S increases. The presumably correct equation is obtained by a derivation analogous to that previously given for absorption of excitation:

$$C = 10^{A'g}[2.3A'S/(1 - 10^{-A'S})] \qquad (5.3)$$

Essentially the same equation, but in terms of transmittance, is also derived in a recent paper by Christmann et al. [37], who thus corrected the fluorescence measured from a quinine sulphate solution to within 3 per cent of the undepressed value in solutions containing fluorescein added to give absorbances up to 2.0 cm^{-1}. The cell parameters used were derived from the experimental data, but agreed to within about 10 per cent of the nominal values. (This corresponds to a difference of 20 per cent in correction factors for the high absorbance regions examined.) The simple 10^{Ad} factor also gave satisfactorily corrected results for absorbances up to 1.0 cm^{-1}, whereafter an increasing over-correction occurred.

When the inner filter absorption of fluorescence is contributed to by the fluorophore itself, secondary fluorescence emission occurs; hence, the effective absorbance of the solution will be less than the measured value, and the calculated correction factor will be an overestimate. This explains [38] the observation that the average molar absorptivity for reabsorption of fluorescence is proportional to (concentration × pathlength)$^{-\frac{1}{2}}$.

(d) *In line geometry*
Now, the fluorescence collected may be emitted throughout the whole depth of the solution. Thus, in equation (5.2), $d_1 = 0$, and d_2 becomes the cell depth; and in the absence of absorption of fluorescence:

$$C = 2.3Ad_2/(1 - 10^{-Ad_2}) \qquad (5.4)$$

If the fluorescence is absorbed:

$$C = 2.3d_2\,(A'-A)/\,10^{-Ad_2} - 10^{-A'd_2})$$

When the absorbance is due only to the fluorophore, the observed fluorescence should increase linearly with $1 - 10^{-Ad_2}$ (because

$2.3Ad_2/(1 - 10^{-Ad_2}) = F_0/F$, $F_0 =$ (concentration) $K = AK'$, therefore $F = (1 - 10^{-Ad_2})/(2.3d_2 K')$; where the Ks are constants). This is confirmed by Milkey and Fletcher [39] for values of Ad_2 up to about 0.9. There seems to be no data published from experiments made with mixtures by which the validity of equation (5.5) can be adequately tested.

(e) Front surface geometry

Because the effective pathlength in this case, with concentrated solutions, is usually assumed to be very small, few attempts seem to have been made to assess the validity of any calculated correction factors. With less concentrated solutions, results from varied pathlengths have been extrapolated to zero pathlengths, but the intercepts have proved difficult to measure because of the nonlinear nature of the extrapolations used [26].

Where the pathlengths of the excitation and the emission within the sample are the same (d), the correction factor at a specified wavelength is:

$$C = 2.3d(A + A')/[1 - 10^{-(A + A')d}] \qquad (5.6)$$

This is essentially an inverted form of Leese and Wehry's [26] *attenuation factor*. With allowance for the spectral distribution of the excitation and of the monitored fluorescence over the monochromators' bandpasses, consistent results are obtained over a considerable concentration range [26], as mentioned previously. Consistent results are also obtained from a similar relationship empirically parameterized with data obtained at varying pathlengths [26].

However, more complex relationships prevail when, for instance, samples differing widely in absorbance are compared. Data from the numerical integration of a relevant equation are tabulated by Mode and Sisson [14], which enable consistent results to be obtained for the quantum yield of quinine sulphate in solutions varying in absorbance from 24.3 to 2.43.

5.1.5 Absorbances and pathlengths for correction factor calculations

(a) Absorbances

Because of the likely differences in the bandwidths and in the spectral distribution of the light sources used in different instruments, it is clearly preferable that the absorbances used in the calculation of correction factors should be determined with the same light beam used for the fluorescence measurements, that is, both sets of measurements should be made in the fluorescence spectrometer. In some instrumentation, the two sets can be made simultaneously [34] (the fluorescence signal can be expressed with reference to the light transmitted by the sample [40], but this does not provide a true correction for inner filter effects), but even without this important refinement, absorbances can be measured with most instruments by monitoring the light transmitted by a sample with a magnesium oxide screen or a quantum counter. In the latter case, low results may be obtained for highly absorbing solutions if they are strongly fluorescent [41].

(b) *Pathlengths*

All correction factor expressions require that the excitation and emission pathlengths, under the conditions within the sample cell, are known. For many instruments, pathlengths cannot be accurately defined — for instance, the excitation beam may not be parallel-sided, and the emission may be collected from an undefined area of the cell surface. Apart from this problem, some of the signal will be due to reflected excitation and emission [42], which increases apparent pathlengths. (The consequences of this effect can be reduced by the use of blackening on irrelevant cell surfaces [12]).

For relatively dilute solutions, the correction factors are not particularly sensitive to error in pathlength values [31], but generally it is a useful expedient to treat pathlength values as empirical parameters, analogously to Holland *et al.* [29]. In the simplest case, for example, if $F_0 = 10^{(A + A')d}$ F and $F_0 = K(\text{conc.})$ hold, then d is the negative slope of a log $(F/\text{conc.})$ against $(A + A')$ plot.

5.1.6 *Experimental countermeasures*

Under the circumstances mentioned before, when a solution cannot be diluted, various techniques are available that reduce the influence of inner filter effects:

(a) Provided that an adequate signal strength can be obtained, an excitation wavelength is selected in regions of low absorbance [20, 27]; for instance, at the long-wave extremity of the main excitation band. The technique is valuable, given that varying levels of extraneous absorption are absent.

(b) In the comparison of samples, their concentrations are adjusted so that their absorbances are equal at the relevant wavelengths [1]. The addition of a non-luminescent absorber to give matched absorbances has been used in a phosphorescence [43] as well as in a fluorescence application [31]. But if generally applied, this could well introduce more problems than it might solve (for example, intermolecular excited state processes, heavy atom quenching).

(c) Excitation and emission wavelengths may be chosen in order that the effects of variation in absorbance at the two wavelengths offset one another. Data tabulated to facilitate the choice are given by van Slegeren et al. [44], but the occasions when the idea is likely to be useful are infrequent.

(d) When an analysis involves the formation and determination of a fluorescent derivative by means of an absorbing, non-fluorescent reagent, excitation wavelengths may be selected where the absorbance due to the reagent and the derivative are equal [44].

(e) In perpendicular geometry, pathlengths may be minimized by the use of short pathlength (micro) cells [20, 25], or of cell spacers to offset the position of the cell so that the intersection of the excitation and emission optical axes is brought close to the cell edge enclosed by the axes [20, 44]. With this arrangement, the position of the cell, and the slit widths used, require careful adjustment if scattering effects from the cell walls are to be kept at a low level.

(f) Frontal illumination may be used, but with deference to the already mentioned limitations of the techniques.

(g) In many analyses, a fluorescent derivative of the analyte is formed that can often be separated by HPLC. Relative to the cost of a fluorescence spectrometer, the additional cost of fabrication of a chromatograph is insignificant. By this technique, interfering chromophores may be separated, and, in the capillary cells used, inner filter effects are in any case small.

5.2 Sample cells

Cells for fluorescence spectrometry must in general meet criteria similar to those in UV absorption spectrometry, which are covered by the terms of reference of a recent working party [45]. The following points, however, seem especially pertinent here.

5.2.1 Cell types

(a) *Cylindrical cells*
The term is used ambiguously in the luminescence literature. Such cells may be either discs, with a pair of optically flat faces, or test tube-shaped. The former design is used where samples are to be degassed, and is most suited to in line or frontal geometry. In the perpendicular arrangement, the emitted light is collected through the curved side of the disc. In test tube-shaped cells, which in an elongated form are generally used for low temperature work, both excitation and emission must pass through curved cell walls, where the angle of incidence varies from point to point. Hence, an accurate estimate of pathlengths is difficult to make, and, in photochemical work, the effect has given rise to anomalous quantum yields [46]. Cells with plane faces are obviously to be preferred in most fluorescence studies, a low temperature example being provided by Parker and Hatchard [47]. (The value of such a cell would of course be lost if used in the test tube-shaped Dewars normally sold with fluorescence spectrometers.)

Until recently, micro cells and flow cells for HPLC have been tubular-shaped, but a square-sectioned cell holding 20 μl is now available [48] that should improve considerably the quality and precision of spectral data available from micro samples.

(b) *Reflecting cell surfaces*
If irrelevant cell surfaces are converted to reflectors, excitation and emission can be redirected in order to increase the fluorescence intensity available for detection [44, 49, 50]. However, the technique drastically increases the amount of scattered radiation reaching the detector, so that the overall effect on detection limits is usually small, and the spectra are particularly subject to inner filter effects because of the increases in effective pathlengths.

5.2.2 Luminescence of cell materials

In conventional fluorescence spectrometry with perpendicular illumination, the pathlength of the excitation and of the emission is determined by the geometry of the cell holder, and not by the cell dimensions.

No part of the illuminated cell surface is seen by the detector, consequently sensitivity is normally limited by reagent blanks, fluorescent contaminants and instrumental factors, rather than by any luminescence from the cell material. Under some circumstances, however, a considerable interference from the cell material may occur, and an approximately 10^5 variation has been found between the luminescence intensity from optical quality fused quartz (the most highly luminescent) and fused synthetic silica [1]. The interference can be important in frontal geometry, where the irradiated cell surface is viewed by the detector. Thus, an apparently delayed fluorescence may under some circumstances originate in phosphorescence from the cell material [51]. This depends on the material of construction of the cell and is most pronounced with frontal illumination.

The presence of finely dispersed material in solution can increase the cell wall luminescence seen [52], but the most serious interference occurs in spectra from samples in micro cells, for instance those used for HPLC, where the volume of the illuminated silica seen by the detector is often greater than that of the solution. The effect may be reduced by the square section cell [48], and by careful alignment of the illuminated part of the cell in the sample compartment. Even so, in some wavelength regions, sensitivity is very much restricted by the silica luminescence, although high purity silica may be used.

The silica luminescence is also a restricting factor in low temperature work with capillary cells, although the use of polarizer reduces the interference [53]; and the interference is compounded by the contribution from the walls of Dewar flasks, though these can be dispensed with [54]. Other techniques that are likely to be restricted in sensitivity are the use of silica capillary cells in fluoromicroscopy [55] and the use of silica elements in multiple internal fluorescence spectroscopy [56]. Quite clearly, there are a number of areas where silica luminescence will be a limiting factor. It is unfortunate, therefore, that there are no generally accepted standards against which the level of this interference can be controlled. Cells in glass and plastics materials are available, which are relatively robust and cheap. But they are of use only when measurements at wavelengths greater than about 320 nm are needed.

5.2.3 *Windowless and other cell types*

(a) *Windowless cells*
Sometimes silica windows have been eliminated altogether. McHard et al. [57] describe a cell for front surface illumination in which the

sample is retained by surface tension; and for microscope work, Rutili et al. [58] present samples of 2–10 nl in a haemacytometer chamber. But such techniques permit only very limited control over the sample environment. A valuable development is described by Smith et al. [59], in which a flow cell is illuminated axially in a front surface mode by means of a bifurcated glass fibre optic. Presumably a major obstacle in the development of the technique may be the presence of background luminescence from the fibre optic material.

(b) *Other cell assemblies*

The precision and sensitivity of much fluorescence work is often improved when sample solutions are deoxygenated. This suppresses oxygen quenching of fluorescence and photo-oxidation of fluorophores. Standard square-sectioned cells can be readily fitted up, for nitrogen-purging, with inlet and outlet lines [60] through which not only nitrogen but also samples can be transferred. Because the transfer of samples is now effected more cleanly and rapidly, total analysis times are not much affected by the time taken for the purging, which in small cells can be completed in about a minute.

A variable pathlength arrangement for perpendicular illumination has been described recently [21] which is used in the exploitation of inner filter effects for the determination of quantum yields. A similar idea had been mooted before [19]. The minimization of inner filter effects is facilitated by the reduction of cell pathlengths below the conventional 0.5 cm. When only small amounts of sample are available this is best done with fixed short pathlength cells, of low volume. Only a few sizes are at present readily available.

5.3 Recommendations

(a) The absorption spectrum of a solution should invariably be determined prior to any fluorescence measurements on it. Any published fluorescence spectra should be accompanied by their corresponding absorption spectra, which should be recorded with reference to a blank that is entirely transparent over the spectral range of interest.

(b) To minimize inner filter effects solutions should be diluted as far as the requirements of accuracy permit. Correction factors should always be calculated. Whether or not they are significant can only be decided when their magnitude is known, and with reference to the case concerned.

(c) The applicability of any particular relationship for the calculation of correction factors will vary according to the instrumentation used, as well as to the cell geometry, and must be empirically confirmed and empirically parameterized if necessary.

(d) There is a considerable potential usage for spectrometers on which fluorescence spectra can be recorded simultaneously with absorption spectra, and simultaneously corrected for inner filter effects.

(e) Standards for fluorescent impurities in fused silica should be established.

(f) Improved fluorimetric accuracy would result if deoxygenation assemblies and a larger range of cell sizes were in more widespread use.

References

1. Parker, C. A. and Rees, W. T. (1962), *Analyst*, **87**, 83.
2. Livingstone, R. (1957), *J. Phys. Chem.*, **61**, 860.
3. Stokes, G. G. (1852), *Phil. Trans.*, **142**, 463.
4. Chapman, J. H., Forster, Th., Kortum, G., Parker, C. A., Lippert, E., Melhuish, W. H. (1963), *Appl. Spectros.*, **17**, 171.
5. Parker, C. A. (1968), *Photoluminescence of Solutions*, Elsevier, p. 220.
6. Udenfriend, S. (1969), *Fluorescence Assay in Biology and Medicine*, Academic Press, **2**, 184.
7. Guilbault, G. G. (1973), *Practical Fluorescence*, Marcel Dekker, Inc., p. 18.
8. Brandt, R., Olsen, M. J., Cheronis, N. D. (1963), *Science*, **139**, 1063.
9. Lloyd, J. B. F., (1971), *J. Forensic Sci. Soc.*, **11**, 529.
10. Birks, J. B. and Munro, I. H. (1967), *Progr. Reaction Kinetics*, **4**, 239.
11. Birks, J. B. (1976), *J. Research*, NBS, **80A**, 389.
12. Melhuish, W. H. (1961), *J. Phys. Chem.*, **65**, 229.
13. Melhuish, W. H. (1972), *J. Research*, NBS, **76A**, 547.
14. Mode, V. A. and Sisson, D. H. (1974), *Anal. Chem.*, **46**, 200.
15. Bierzynski, A., and Jansy, J., (1974/5) *J. Photochem.*, **3**, 431.
16. Fletcher, M. H. (1963), *Anal. Chem.*, **35**, 278.
17. Fletcher, M. H. (1963), *Anal. Chem.*, **35**, 288.
18. Huke, F. B., Heidel, R. H. and Fassel, V. A. (1953), *J. Opt. Soc. Am.*, **43**, 400.
19. Ohnesorge, W. E. and Rogers, L. B. (1959), *Spectrochim. Acta*, **15**, 27.
20. Chen, R. F. and Hayes, J. E. (1965), *Anal. Biochem.*, **13**, 523.
21. Britten, A., Archer-Hall, J. and Lockwood G. (1978), *Analyst*, **103**, 928.
22. Lipsett, F. R. (1967), *Progr. Dielectrics*, **7**, 217.
23. Demas, J. N. and Crosby, G. A. (1971), *J. Phys. Chem.*, **75**, 991.
24. Weill, G. and Calvin, M. (1963), *Biopolymers*, **1**, 401.
25. Chen, R. F. (1967), *Anal. Biochem.*, **19**, 374.
26. Leese, R. A. and Wehry, E. L. (1978), *Anal. Chem.*, **50**, 1193.

27 Parker, C. A. and Barnes, W. J. (1957), *Analyst*, **82**, 606.
28 Ohnesorge, W. E. and Rogers, L. B. (1959), *Spectrochim. Acta*, **15**, 27.
29 Holland, J. F., Teets, R. E., Kelly, P. M. and Timnick, A. (1977), *Anal. Chem.*, **49**, 706.
30 Brand, L. and Witholt, B. (1967), *Methods in Enzymology*, Ed. C. H. W. Hirs, Academic Press, **11**, 776.
31 Franzen, J. S., Kuo, I. and Chung, A. E. (1972), *Anal. Biochem.*, **47**, 426.
32 Parker, C. A. (1968), *Photoluminescence of Solutions*, Elsevier p. 222.
33 Gill, J. E. (1970), *Appl. Spectros.*, **24**, 588.
34 Holland, J. F., Teets, R. S. and Timnick, A. (1973), *Anal. Chem.*, **45**, 145.
35 Mertens, M. L., and Kagi, J. H. R. (1979), *Anal. Biochem.*, **96**, 448.
36 Wu, F. Y.-H., Tu, S.-C., Wu, C.-W. and McCormick, D. B. (1970), *Biochem. Biophys. Rsch. Com.*, **41**, 381.
37 Christmann, D. R., Crouch, S. R., Holland, J. F., and Timnick, A. (1980), *Anal. Chem.* **52**, 291.
38 Rohatgi, K. K. and Singhal, G. S. (1962), *Anal. Chem.*, **34**, 1702.
39 Milkey, R. G. and Fletcher, M. H. (1957), *J. Am. Chem. Soc.*, **79**, 5425.
40 Turner, G. K. (1964), *Science*, **146**, 183.
41 Mehler, A. H., Bloom, B., Ahrendt, M. E. and Stetten, D. (1957), *Science*, **126**, 1285.
42 Shurcliff, W. A. and Jones, R. C. (1949), *J. Opt. Soc. Am.*, **39**, 912.
43 Zander, M. (1968), *Phosphorimetry*, Academic Press, p. 136.
44 van Slageren, R., Den Boeff, G. and van der Linden, W. E. (1973), *Talanta*, **20**, 501.
45 UV Spectrometry Group (1981), *Standards in Absorption Spectrometry*, Eds. C. Burgess and A. Knowles, Chapman and Hall.
46 Vesley, G. F. (1971), *Mol. Photochem.*, **3**, 193.
47 Parker, C. A. and Hatchard, C. G. (1962), *Analyst*, **87**, 664.
48 Dicesare, J. L. and Stoveken, J. (1977), *Chromatogr. Newsl.*, **5**, 31.
49 Braunsberg, H. and Osborn, S. B. (1952), *Anal. Chim. Acta*, **6**, 84.
50 Laikin, M. (1963), *Appl. Spectros.*, **17**, 26.
51 Parker, C. A. and Joyce, T. A. (1966), *J. Chem. Soc.* (A), 821.
52 Price, J. M. Kaihara, M. and Howerton, H. K. (1962), *Appl. Optics*, **1**, 521.
53 Lukasiewicz, R. J., Rozynes, P. A., Saunders, L. B. and Winefordner, J. D. (1972), *Anal. Chem.*, **44**, 237.
54 Colmsjo, A. and Stenberg, U. (1979), *Anal. Chem.*, **51**, 145.
55 Sernetz, M. and Thaer, A. (1972), *Anal. Biochem.*, **50**, 98.
56 Harrick, N. J. and Loeb, G. I. (1973), *Anal. Chem.*, **45**, 687.
57 McHard, J. A. and Winefordner, J. D. (1972), *Anal. Chem.*, **44**, 1922.
58 Rutili, G., Arfors, K. -E. and Ulfendahl, H. R. (1976), *Anal. Biochem.*, **72**, 539.
59 Smith, R. M., Jackson, K. W. and Aldous, K. M. (1977), *Anal. Chem.*, **49**, 2051.
60 Lloyd, J. B. F. (1974), *Analyst*, **99**, 729.

6 Temperature effects and photodecomposition in fluorescence spectrometry

6.1 Errors caused by temperature effects

The fluorescence efficiency of many compounds is very sensitive to temperature variations and for accurate work, temperature regulation is necessary [1, 2]. Fluorescence yields and decay times usually decrease with increasing temperature due to enhancement of the probability for internal conversion and intersystem crossing to the triplet manifold. On the other hand, the rate of collisional quenching decreases as the viscosity of the medium is increased so that collisional quenching of fluorescence in liquid media is less serious as the temperature is lowered. (This generalization does not always hold if the solute has two or more excited states which are slightly different in energy. In a compound in which T_1^* lies just below S_1^* e.g. anthraquinone [3] intersystem crossing from S_1^* to T_1^* may be followed by thermal excitation of the triplet back to S_1^* - hence the fluorescence intensity will increase with temperature).

There have been numerous studies of the effect of temperature on the intensity of fluorescence and the reader is referred elsewhere for representative publications [4–7]. Over a wide temperature range, the fluorescence quantum yield $\phi_f(T)$ as a function of temperature is determined relative to the value ϕ_f° at room temperature by the relation:

$$\phi_f(T) = \phi_f^\circ \frac{\int_0^\infty F(\bar{\nu}, T)\, d\bar{\nu}}{\int_0^\infty F_0(\bar{\nu})\, d\bar{\nu}} \cdot \frac{A_0}{A(T)} \cdot \frac{n^2(T)}{n_0^2}$$

where F corresponds to the corrected fluorescence spectrum in terms of the relative number of quanta per unit wavenumber, A is the absorbance of the sample at the excitation wavelength and n is the refractive index of the solvent. The proper use of this equation at room temperature has been discussed [8]. The fluorescence quantum yield at room temperature ϕ_f° is independently established relative to a known standard (see Chapter 8). The absorbance A changes with temperature and this change requires measurement of the absorption spectrum over the same range for $\phi_f(T)$ determinations [5]. The refractive index term corrects for the temperature variation in the angles of the emerging emission rays from the plane of the cuvette–air interface. Recent work [9] has questioned the n^2 correction factor since it was shown that proper correction is a strong function of the geometry of the sample compartment. Another effect of n of more fundamental importance is its influence on the intrinsic radiative parameters of the excited state. Studies with diphenyl-anthracence (in isopentane at temperatures down to -160° C) revealed a decrease in fluorescence lifetime consistent with an n^{-2} to n^{-3} dependence [10].

Over the more practical temperature range from 15 to 30° C, changes in fluorescence intensity of 1–5%°C^{-1} may be encountered [2] (see Fig. 6.1).

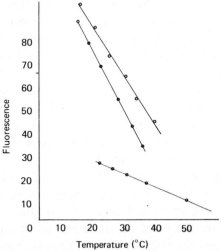

Fig. 6.1 *Variations in the fluorescence intensity of several compounds as a function of temperature. All compounds were dissolved in 0.1 M phosphate buffer, pH 7.0 except quinine [2].* ● *tryptophan or indoleacetic acid,* ○ *indoleacetic acid in buffer saturated with benzene,* ■ *quinine in 0.1 N sulphuric acid.*

6.2 Countermeasures and recommendations for temperature effects

The substantial variations in fluorescence intensity with temperature clearly demonstrate that temperature control should be exercised for maximum precision and accuracy. Constant temperature cell blocks are available for most fluorescence spectrometers and should be used in most instances: a temperature variation of not more than ±0.1° C is desirable. In cases where samples must be stored under refrigeration, care must be taken to bring the sample back to room temperature before measurement.

6.3 Errors caused by photolysis effects

Ultraviolet radiation commonly required for fluorescence excitation of many compounds in solution may also induce photochemical decomposition leading to potential errors in fluorescence measurements. For weakly fluorescing and/or dilute solutions, it is tempting to increase the excitation intensity (usually by opening the slits of the excitation monochromator) so that a signal or recorder trace with an adequate photometric signal-to-noise ratio is obtained. This effect is commonly encountered by users attempting to check the manufacturer's specification for the detectability of a very dilute quinine sulphate solution (see Section 4.2 above): with many fluorescence spectrometers it is necessary to work at the highest excitation intensity in order to obtain a quinine sulphate signal significantly different from that of the blank solvent. Many users must have noticed how the fluorescence signal appears to decrease rapidly with time of irradiation in these circumstances.

Parker [11] has pointed out the two kinds of error arising from photolysis in fluorescence spectrometry. The first, and most common effect, is simple decomposition of a fluorescent solute during the time taken to make a measurement. This often leads to a reduction in fluorescence intensity with time. The second kind of error arises when the reagent used to form a fluorescent compound undergoes photolysis to produce another fluorescent product. In both cases, photolysis of all but the most sensitive solutions can be prevented or minimized by using the following commonsense measures.

6.4 Countermeasures and recommendations

(a) Always use the longest feasible wavelength for excitation.
(b) Measure the fluorescence intensity immediately after

excitation has commenced and use a shutter to minimize the irradiation time.

(c) Protect unstable solutions from ambient light such as the sun or artificial lights (especially fluorescent lamps) by storing in a dark bottle*.

(d) Deoxygenation of the solution often helps to minimize photodecomposition.

The effect of photodecomposition is much less in concentrated than in dilute solutions but the inner filter effect and concentration quenching may then impose other practical limits (see Section 5.1). Improvements in instrument sensitivity, particularly by the application of photon counting or lock-in amplification, are required as a last resort to detect weakly fluorescing and dilute solutions without any accompanying photodecomposition.

If a solute in a cuvette is stirred during measurement, the rate of decomposition may be reduced but this is a considerable problem with microcuvettes and in the flow cells used with HPLC detectors. Stirring can only produce a marginal improvement in the case of new fluorophor production since a continuously rising blank reading sets a limit on the sensitivity of the method.

For precise work, considerable care has to be taken to verify that a decrease in fluorescence intensity with time is not an instrumental artifact caused, for example, by a change in lamp intensity, solvent temperature or photodetector sensitivity.

Very few fluorescence spectrometers show drifts in sensitivity with time of less than a few percent per hour. A convenient *drift base line* can be determined by monitoring stray or scattered light in the absence of any fluorophor (see Section 3.2 above) or by using a known stable solution, for example, *p*-terphenyl in liquid or solid solutions.

Notwithstanding the problems of photochemical decomposition in many studies, it is also noteworthy that a method has been described that uses such decomposition in the analytical determination of various compounds [13]. In dilute solutions, both the rate of photochemical reaction and the fluorescent intensity are linearly dependent on the incident light intensity. Pseudo first-order decay curves of fluorescence signal versus time are thus obtained and initial

---

*Fluorescent lamps have been shown to have a dramatic effect on the stability of 1,10-phenanthroline solutions which are commonly used in the ferrioxalate chemical actinometer [12].

fluorophor concentrations can be measured either by extrapolation of the initial rate of decay or by digital integration of the fluorescence signal for a fixed time.

References

1. Parker, C. A. (1968), *Photoluminescence of solutions*, Elsevier, Amsterdam, p. 81.
2. Guilbault, G. G. (1973), *Practical fluorescence, theory, methods and techniques,* Marcel Dekker, New York, pp. 24, 123.
3. Carlson, S. A. and Hercules, D. M. (1971), *J. Amer. Chem. Soc.*, **93**, 5611.
4. Cehelnik, E. D., Cundall, R. B., Lockwood, J. R. and Palmer, T. F. (1975), *J. Phys. Chem.*, **79**, 1369.
5. Mantulin, W. W. and Huber, J. R. (1973), *Photochem. Photobiol.,* **17**, 139.
6. Suzuki, S., Fujii, T., Imai, A. and Akahori, H. (1977), *J. Phys. Chem.*, **81**, 1592.
7. Bowen, E. J. and Sahu, J. (1959), *J. Phys. Chem.*, **63**, 4.
8. Demas, J. N. and Crosby, G. A. (1971), *J. Phys. Chem.*, **75**, 991.
9. Morris, J. V., Mahaney, M. A. and Huber, J. R. (1976), *J. Phys. Chem.*, **80**, 969.
10. Olmsted, J. (1976), *Chem. Phys. Letters*, **38**, 287.
11. Parker, C. A. (1968), *Photoluminescence of solutions*, Elsevier, Amsterdam, p. 426.
12. Bowman, W. D. and Demas, J. N. (1976), *J. Phys. Chem.*, **80**, 2434.
13. Lukasiewicz, R. J. and Fitzgerald, J. M. (1973), *Analyt. Chem.*, **45**, 511.

7 Correction of excitation and emission spectra

7.1 Introduction: the need for correction procedures

In order to determine the fluorescence characteristics of a molecule free from instrumental artefacts, it is necessary to correct the observed spectra for the wavelength-dependent efficiency of the excitation source and the detector system. These corrections have been advocated widely for twenty years but are far from universally performed.

They are essential if spectra obtained on different instruments are to be compared, and for any study of the fundamental fluorescence characteristics of a molecule, including its quantum yield. On the other hand, many simple analytical applications, and many empirical applications in biochemistry, do not require corrections for instrumental response functions. However, it is probably true to say that more information could be extracted from many of the biochemical experiments if corrected spectra were obtained.

The procedures involved in obtaining corrected spectra have been reviewed by, among others, Parker [1], Demas and Crosby [2] and Melhuish [3].

7.2 Excitation spectra

Corrections of excitation spectra involve the determination of the relative photon spectral irradiance (relative numbers of quanta per second per unit area per unit bandwidth) of the excitation system (lamp and monochromator) as a function of wavelength. This requires the use of a detector whose response function is accurately known, and preferably flat. The two most commonly used such detectors are thermal detectors and quantum counters.

7.2.1 Thermal detectors

These are of two types, thermopiles and the more recent *pyroelectric detectors* which depend upon the high temperature coefficient of the spontaneous polarization of a ferroelectric. Both detectors are coated with a layer which absorbs the incident radiation and thus leads to heating of the detector. Provided that this coating has a reflectivity which is independent of wavelength, the detector should give an equal output for an equal amount of energy, irrespective of wavelength. (Note that since the response of a thermal detector is proportional to *energy*, rather than numbers of quanta, it must be multiplied by the wavelength of the incident light to give the desired plot of relative photon spectral irradiance versus wavelength).

Thermal detectors are often *assumed* to have a perfectly flat response function [4, 5]. 'However, the empirical verification of this assumption is a tedious process which is, apparently, usually avoided by manufacturer and user alike' [6].

One detector coating (101−C10 Black, 3M Company) is quoted [7] as having a reflectance of 1.9 per cent at 800 nm rising to 4.9 per cent at 250 nm, which would give a response flat to within 3 per cent over this range, but Christensen and Ames [6] observed substantially more variation of response as a function of wavelength, notably in the ultraviolet.

The wide range of wavelength sensitivity of thermal detectors can be a source of error if the light beam contains any stray infrared radiation. Attempts to remove stray light with a filter may be of no help if the energy absorbed by the filter is re-emitted in the infrared region.

The major difficulty in the use of thermal detectors is, however, their low sensitivity (particularly in the case of thermopiles). This is especially serious in the ultraviolet, where lamp output is low, and when narrow bandwidths are required. (The excitation source must of course be calibrated with the bandwidths to be used for the sample, particularly in view of the marked emission lines near 470 nm observed with xenon arcs; see Section 2.3.)

7.2.2 Quantum counters

The use of suitable fluorescent compounds as quantum counters is based on the early work of Vavilov [8] and Bowen [9] among others. If the absorption spectrum and concentration of the compound are such that all ($>$ 99 per cent) the light in a given wavelength range is absorbed, and if the emission spectrum and quantum yield are independent of excitation wavelength, then the intensity of the fluorescence it emits is simply proportional to the photon spectral irradiance of the excitation source, regardless of wavelength (provided that the transmittance of the quantum counter window is taken into account).

By far the most commonly used compound in quantum counters is rhodamine B (spectra in Appendix), introduced by Melhuish [10, 11]. This is used at concentrations of 3–8 g l^{-1} commonly in glycerol or ethylene glycol, though trifluoroethanol has also been used [12]. A number of experiments [11, 13, 14, 15] have shown that the fluorescence efficiency (at around 610–620 nm) of such a solution is constant to ± 2 per cent in the range of 350–600 nm and to ± 5 per cent down to 250 nm. Recently, the related rhodamine 101 has been used in glycerol solution; this shows less temperature dependence of its fluorescence than does rhodamine B.

Other compounds which have been used as quantum counters (see Appendix for spectra) include sodium 1-dimethylaminonaphthalene-5-sulphonate [16, 17], fluorescein and quinine sulphate. The latter compounds are far from ideal; fluorescein has very low absorbance at 340–360 nm leading to potential errors from trace absorbing impurities [2, 11], while quinine has the same problem at ~270 nm, and in addition the emission spectrum shifts on excitation at wavelengths > 350 nm (see below, Section 7.3). Dimethylaminonaphthalene-5-sulphonate, however, is well suited as a quantum counter in the range 210–400 nm; solutions of 0.01–0.02 M in 0.1 N NaOH or Na_2CO_3 are suitable. Both rhodamine B and 1-dimethylaminonaphthalene-5-sulphonate are commercially available in adequate purity. Drexhage [18] has proposed the carbocyanine dye HIDC as a quantum counter whose useful range extends up to 700 nm.

HIDC

Nile Blue A may also be suitable as a high-wavelength quantum counter [19].

Two geometries have been used with quantum counters: *front face* (a) and *transmission* (b):

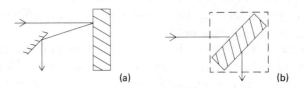

Most commercial instruments have *solid sample* or *front-surface* accessories which can be used for geometry (a). Melhuish [3] recommends geometry (b), used by Parker [20], which uses the full thickness of the quantum counter as a filter. A mushroom-shaped cell has been described [21] which minimizes errors due to variation in the depth of penetration of the excitation light with wavelength, and to polarization effects; however, since the detector is in-line with the excitation beam, stray light must be rigorously excluded.

Care must be taken to eliminate errors due to second order radiation passed by the excitation monochromator; a 350 nm cut-off filter should be placed in the excitation beam when calibrating at > 400 nm [3]. Alternatively, comparison of the calibration curves for 400–600 nm obtained with and without a simple glass filter in the excitation beam can indicate whether second-order radiation is leading to error [17]. Chen [17] also recommends a comparison of the rhodamine B and l-dimethylaminonaphthalene-5-sulphonate quantum counters as a test for errors due to stray light which would have less serious effects on the latter because of its more restricted absorption region. Finally, setting the detection system to view the long-wavelength tail of the emission band of the quantum counter minimizes errors due to reabsorption and shifts in the position of the emission spot [13].

7.2.3 Other methods

(a) *Actinometry*

Chemical actinometry is one of the classical methods of measuring light intensity. The best-known system is the potassium ferrioxalate actinometer [22]: absorbed light reduces ferric to ferrous iron, which is then estimated as the *tris* (1, 10-phenanthroline) complex. The response of the system is not strongly wavelength-dependent and has been well calibrated, but the wavelength range is limited to

250–480 nm (with pathlengths up to 5 cm). The main disadvantage is, as with the thermopile, that low sensitivity exposure times of minutes or hours may be required. This obviously leads to errors if the light source is at all unstable, and since a separate sample must be exposed and worked up for each wavelength, determination of a complete calibration of the excitation system becomes extremely tedious. Other actinometers proposed for longer wavelengths (Reinecke's salt, and the *tris* (2, 2-bipyridyl) osmium (II) – sensitized photo-oxidation of tetramethylethylene) are at least ten-fold less sensitive still.

(b) *Comparison of excitation and absorption spectra*
Since the 'true' excitation spectrum of a fluorophore should be identical to its absorption spectrum, a comparison of the experimental excitation spectrum with the absorption spectrum can be used to calculate the wavelength-dependent correction factors for the output of the excitation system. This method has also been described [2] as the optically dilute quantum counter method. Argauer and White [23] calibrated their excitation system in this way using the aluminium chelate of 2, 2-dihydroxy-1, 1-azonaphthalene-4-sulphonate (Pontachrome Blue Black R) and obtained reasonable agreement with a thermopile calibration. This method is not suitable as a primary calibration. The identity of excitation and absorption spectra only holds in general at infinite dilution and self-absorption would produce serious errors. Appreciable error can be introduced whenever the absorption spectrum goes through deep minima. It is important to determine the absorption and excitation spectra under the same conditions; in particular the bandwidths must be the same (and narrow) and the two monochromators must have properly calibrated wavelength scales. This method is, however, useful as a periodic check.

(c) *Use of a calibrated detector system*
Drushel *et al.* [24] replaced the sample by a MgO screen and used a detector system (monochromator and photomultiplier) previously calibrated against a standard lamp (see below, Section 7.3.1) to measure the relative photon spectral irradiance of the excitation system, by scanning the two monochromators together. Since other methods for calibrating excitation systems, such as the quantum counter, are both easier and more precise than those for calibrating detector systems this method has not been widely used.

7.2.4 Continuous correction

It is clearly desirable to have a method of performing the correction of the excitation source automatically and continuously, so that a corrected excitation spectrum is recorded by the instrument. This can readily be done by sampling the excitation beam and directing a part of it to a thermal detector or a quantum counter/photomultiplier. The output of the sample detector is then ratioed to the output of this reference detector. Such an arrangement has the additional substantial advantage that fluctuations in the output of the lamp are automatically compensated for.

This kind of arrangement is used on a number of commercial instruments. For example, in the Aminco SPF-1000, the excitation beam is sampled by a chopper mirror and directed on to a pyroelectric detector. In the Perkin-Elmer MPF-44, a quartz beam-splitter directs a small part of the excitation beam on to a rhodamine B quantum counter. It is important in such an arrangement to take account of the wavelength-dependence of the reflectivity and transmittance of the beam splitter. The additional correction for this can be determined by placing a quantum-counting screen in the sample position, and determining the ratio of the sample and reference photomultiplier outputs as a function of wavelength. (In the Perkin-Elmer instrument this correction is *programmed* into a multi-tap potentiometer ganged to the excitation monochromator.) To minimize polarization effects, the beam splitter should be positioned so as to make a small angle (about 15°) with the excitation beam.

7.3 Emission spectra

The correction of emission spectra is both more important and more difficult than correction of excitation spectra. In the simplest case, at least, the information in the excitation spectrum is also obtainable from the absorption spectrum, while that in the emission spectrum can be obtained in no other way. A corrected emission spectrum is also required for the determination of quantum yields. Calibration of the emission monochromator and photomultiplier requires the use of either a light source of known spectral distribution or a compound or series of compounds with known true emission spectra. Since the latter can only have been determined using a detector calibrated in some other way, they can only be regarded as secondary standards.

7.3.1 Primary standards

(a) *Standard lamps*

Both the National Bureau of Standards (US) and the National Physical Laboratory (UK) produce calibrated tungsten lamps (for example, to give a given colour temperature when operated with a given applied voltage) and Melhuish [3] regards these as the most accurate spectral radiant sources available, the uncertainty in output ranging from 3 per cent at long wavelength to 8 per cent at short wavelengths. Recent work (J. Moore in a private communication) suggests that these errors can be reduced to about 1 per cent. (Once the colour temperatue is known, the light output can be calculated from the black-body radiation law, the emissivity of tungsten, and the transmission of the lamp window.) The precautions required in the use of such a standard lamp for calibration purposes have been outlined by Christensen and Ames [6]; considerable care is needed to ensure that the conditions of use are those for which the calibration is valid.

A practical difficulty in the use of standard tungsten lamps is the intensity of the light produced in the visible region. The lamp must either be placed at some distance from the instrument (Chen [17] placed his lamp at the far end of a 16 ft cardboard tube), or some kind of attenuator must be used. The commonest kind of neutral attenuator used is the blackened copper mesh or grid, though some care is required to place such an attenuator in the beam in a reproducible fashion. Recently, the National Bureau of Standards (US) and the National Physical Laboratory (UK) have been investigating neutral filters made of a film of Inconel on a quartz plate, which appear to have a transmittance which is essentially independent of wavelength [25]. Allowance must also be made for the wavelength-dependence of reflectivity of any mirror or scatterer used to direct the lamp beam into the emission monochromator (see section below).

In contrast, many tungsten lamps produce too little light in the ultraviolet to be useful for calibration, since stray light from the intense visible output interferes too much. Thus Chen [17] and Melhuish [3] reported that their standard lamps were only usable down to 400 nm and 340 nm, respectively. More recently, however, lamps calibrated down to 300 nm and usable at still shorter wavelengths have become available [26, 27].

(b) *Use of a calibrated excitation source*

Since the output of a xenon lamp/monochromator system can fairly readily be calibrated by means of a quantum counter, (Section 9.2.2), it can then be used as a light source of known spectral distribution. The simplicity of this method has led to its widespread use since it was introduced in 1962 [11, 28]. Calibration curves obtained by this method have been found to agree very well with those obtained by use of a standard lamp [11, 17, 28], though the sharp emission lines of xenon lamps at ~ 470 nm make very precise corrections in this region difficult. Since it is almost impossible to scan both emission and excitation monochromators simultaneously so that they pass precisely the same wavelength, the excitation monochromator should be set at 5–10 nm intervals and the emission monochromator scanned through the peak so that the peak height (or, ideally, the peak area) can be measured; the bandwidth of the emission monochromator should be 4–5 times greater than that of the excitation monochromator [3].

The light from the excitation system must be directed into the detector by a reflector or scatterer placed in the sample position, and it is obviously important that this should have a minimal dependence of reflectivity on wavelength. Silvered and, particularly, aluminized mirrors have a marked wavelength-dependence of reflectivity. (The use of a mirror is also undesirable from the point of view of polarization effects; see Section 7.4 below). A layer of fresh MgO approx 1 mm thick has the best spectral properties, but its performance degrades markedly on aging or exposure to the atmosphere or ultraviolet light [29]. A number of commercially available materials appear to provide stable surfaces with substantially wavelength-independent reflectivity. These include $BaSO_4$ with K_2SO_4 binder (Eastman White Reflectance Standard; flat to 1 per cent 750–400 nm, and to 5 per cent down to 250 nm [30]) and Halon (a fluorinated polyhydrocarbon; flat to 3 per cent down to 300 nm, and to 6 per cent down to 200 nm [31]). If the excitation system contains a lens, its image size will vary with wavelength, and this could lead to a variation in the fraction of the excitation light seen by the emission monochromator. Borreson and Parker [32] found that this effect could significantly (~ 20%) alter the correction factors.

7.3.2 Secondary standards

(a) Characteristics required of a standard compound
An ideal fluorescence emission standard should satisfy the following criteria:

(i) A broad emission spectrum with no fine structure.
(ii) A large Stokes' shift, leading to a small overlap between absorption and emission spectra and hence to minimal problems of self-absorption.
(iii) A relatively high quantum yield.
(iv) An emission spectrum independent of excitation wavelength (both in shape and quantum yield).
(v) Minimal susceptibility to oxygen quenching.
(vi) Minimal concentration quenching (and excimer formation).
(vii) A completely isotropic (depolarized) emission (see Section 7.4 below).
(viii) It should be readily obtainable in a very pure state and be chemically and photochemically stable.

These criteria (which are essentially the same as those for an ideal quantum yield standard — see Chapter 8) are clearly rather stringent — particularly since several such compounds will be required to cover the wavelength range (approximately 300–700 nm).

In fact criteria (v) and (vii) tend to be mutually exclusive: if the excited state has a long lifetime, there is a greater chance of molecular tumbling averaging out the anisotropy of emission, but also a greater chance of oxygen quenching.

By far the most widely used emission standard is quinine sulphate, but there has been a widespread controversy regarding its precise quantum yield and its suitability as an emission standard.

(b) Quinine sulphate
Melhuish [3] has compared eleven determinations of the corrected emission spectrum of quinine in sulphuric acid. There is a substantial spread in the reported values (normalizing to 1.00 at 2.2×10^3 cm^{-1}, the values at 1.9×10^3 cm^{-1} range from 0.35 to 0.51) which he ascribes to calibration errors, noting his own observations that the spectrum is independent of (i) sulphuric acid concentration (0.1–2 N), (ii) excitation wavelength (260–390 nm) and (iii) the source of the quinine.

However, others have reported substantial changes in the

fluorescence characteristics of quinine depending on these three variables. The details of the many reports bearing on this subject are given in [2] and [33]; the current position seems to be as follows [2, 33–35]:

(i) The quantum yield of quinine increases by 6–7% on increasing the sulphuric acid concentration from 0.1 to 1 N. In perchloric acid, however, quinine fluorescence appears to be independent of acid concentration [36];

(ii) The emission spectrum shifts to longer wavelengths as the excitation wavelength is increased beyond 350 nm. Up to a 15 nm shift has been reported, although the quantum yield does not change.

(iii) Quinine sulphate samples from a number of commercial sources appear to be equivalent, provided that they are carefully dried.

Although quinine has a number of the desirable characteristics of a standard, such as a broad emission spectrum, and minimal susceptibility to oxygen quenching, it thus also has undesirable properties. In addition, its quantum yield is temperature-dependent ($-0.25\%°$ C^{-1}), it is sensitive to halogen quenching (so that only the highest purity sulphuric acid should be used), and although it is chemically stable over periods of months, it shows some photochemical degradation (for example [37]).

(c) *Other proposed standards* (See Appendix for spectra)
2-Naphthol in aqueous buffer [3, 23, 38], shows an emission spectrum which depends markedly on temperature, pH, buffer concentration and 2-naphthol concentration, and is also somewhat photochemically unstable [39]. It is thus unsuitable as a standard.

3-aminophthalimide in H_2SO_4 [3, 23, 34, 38] has an emission maximum at approx. 510 nm. Reasonable agreement (except on the long-wavelength edge of the emission band) exists on the spectrum of this compound. It shows minimal overlap of absorption and emission spectra, and the emission spectrum is independent of excitation wavelength, but its quantum yield (and absorption spectrum) is dependent on the acid concentration [23].

m-Nitrodimethylaniline [3, 23, 34, 38], with an emission maximum at ~ 540 nm, has too low a quantum yield for convenient use as a standard.

2-aminopyridine in H_2SO_4 [3, 40]; there is good agreement on the emission spectrum of this compound but its suitability as a standard has not been thoroughly investigated.

Eosin in 0.1 M NaOH [2, 41] is chemically unstable, and thus unsuitable.

Fluorescein in 0.1 M NaOH or bicarbonate buffer, pH 9.6 [2, 16, 41] has been widely used as a quantum yield standard (Chapter 8). It is not oxygen quenched, but does slowly decompose in solution. There is marked overlap of the absorption and emission spectra, and the high extinction coefficient ($\epsilon \sim 8 \times 10^4 \text{l mol}^{-1} \text{ cm}^{-1}$) of the long-wavelength absorption band leads to serious self-absorption at concentrations of 10^{-6} M and above [42]. The much lower extinction coefficients at shorter wavelengths (~ 350 nm) mean that results obtained with excitation in this region are very susceptible to errors arising from absorbing impurities. Fluorescein is thus unsuitable as a standard.

Anthracene in 95 per cent ethanol [2, 16, 24, 39, 41] also has a number of undesirable properties. There is considerable overlap of absorption and emission spectra, the emission spectrum is highly structured and strongly quenched by oxygen, and purification is not always easy.

Tryptophan in aqueous solution [2, 33]. This is in principle ideal as a standard for many biochemical applications, since the fluorescence of most proteins arises from their tryptophan residues. After some earlier controversy, the value of 0.12 for its quantum yield seems now to be accepted, although corrected emission spectra have not, apparently, been published in numerical form. N-acetyltryptophan amide might be preferable, since its fluorescence is pH-independent. The disadvantages of tryptophan are photochemical instability, overlap of absorption and emission bands and marked temperature-dependence. Bacterial growth limits the long-term stability of aqueous solutions.

9, 10-diphenylanthracene in cyclohexane [2, 39] was used as a quantum yield standard by Berlman; Melhuish [3] found a quantum yield of 0.83 rather than Berlman's (somewhat arbitrary) value of 1.00, but Ware [43] has recently reported a quantum yield of 1.0. The emission band is moderately structured, and significantly overlaps the absorption band, so there is significant reabsorption of fluorescence [35]; the extinction coefficient at 260 nm is however high, so that this can be minimized by dilution. The fluorescence is fairly sensitive to oxygen quenching (the integrated emission intensity of a de-aerated solution being 40 per cent higher than an air-saturated solution [39]).

Phenanthrene has been recommended [35] since it shows no self-quenching and minimal self-absorption. However, it has a highly structured emission band, and shows marked oxygen quenching [39].

Rhodamine B [2, 37, 39] has the advantage that its emission (~ 590 nm) extends to long wavelengths, but the disadvantage that there is considerable overlap of absorption and emission spectra.

p-Terphenyl [37, 39] has a high quantum yield, an emission spectrum centred on 340 nm (in cyclohexane) with only slight structure and very little overlap with the absorption spectrum, and minimal oxygen quenching.

PBD [39], used in liquid scintillation counting, has a similar favourable combination of properties, its emission band being centred at 360 nm.

PBD

1, 1, 4, 4-tetraphenylbutadiene [37, 39] has a broad and structureless emission band centred on 460 nm, very little overlap with the absorption spectrum, and minimal oxygen quenching.

The last three compounds deserve further investigation as fluorescence emission standards, as do the benzothiazoles and benzoxazoles described by Williams and Heller [44], which have large Stokes' shifts and good photochemical stability.

(d) *Solid standards*

It is clear that a solid fluorescence standard which could be fabricated in the dimensions of a standard cuvette would have a number of advantages over standards which have to be used in liquid solution. The increase in convenience would almost certainly encourage more users to correct their emission spectra. Two kinds of solid samples can be envisaged, and these are briefly discussed below (see also Section 2.5). It should be pointed out, however, that problems of emission anisotropy (see Section 7.4 below) would be expected to be encountered with rigid samples, and there is little information on the size of the errors these introduce.

(i) *Aromatic compounds.* A range of aromatic fluorophores can be prepared in acrylic copolymers to give hard, optically clear and homogenous samples [37, 45]. Compounds that have been examined in this way include pyrene, anthracene, 9, 10-diphenylanthracene, perylene, p-terphenyl, tetraphenylbutadiene and rhodamine B. The photochemical stability of the last four compounds in polymer blocks appears to be excellent [37].

(ii) Doped glasses. Reisfeld [46] has considered in detail the absorption and fluorescence spectra of inorganic glasses doped with heavy metal ions, and their possible utility as standard materials. For the present purposes, the metal ions of interest are Tl^+, Pb^{2+}, Ce^{3+} and Cu^+. Reisfeld [46] shows that these ions doped into phosphate, borax or silicate glasses will between them cover the whole spectral range from 270–600 nm (see Table 9.1).

Table 7.1 *Properties of doped inorganic glasses*

Ion	Matrix	Emission maximum (nm)	Emission half bandwidth (nm)
Tl^+	Phosphate	302	90
Ce^{3+}	Phosphate	334	55
Ce^{3+}	Borax	365	65
Ce^{3+}	Silicate	380	85
Pb^{2+}	Phosphate	390	70
Pb^{2+}	Borax	425	165
Cu^+	Phosphate	445	120

The fluorescence intensity is reported to be linearly related to ion concentration over a thousand-fold range. Provided that their stability characteristics are satisfactory, these glasses would seem to be well suited to be standard samples. Their corrected emission spectra do not, however, seem to have been published in numerical form.

(e) *Continuous correction*

An *on-line* correction of emission spectra is not possible in the same way as it is for excitation spectra. Several commercial instruments have multi-tap potentiometers ganged to the emission monochromator on which the correction function is set up. With the increasing use of microprocessors as an integral part of instruments, the emission correction should become very much easier, and spectral correction devices using microprocessors are now commercially available (see Appendix).

7.4 Polarization effects

These are of two kinds: errors arising from the fact that grating monochromators transmit differently for light polarized parallel and perpendicular to the slits, so that the excitation light is not unpolarized and the detection system is not unbiased; and those coming from an anisotropic emission of the fluorophore. Both these can effect the light intensity seen by the detector and, to the extent that they vary with wavelength, can alter the correction factors for the emission spectrum. In addition, the first of these is the origin of the instrumental correction required in measurements of polarization.

7.4.1 Instrumental effects

If the emission monochromator and detector have been calibrated with a standard lamp, or with a calibrated excitation system and a scatterer (Section 7.3), the response function obtained is essentially that to unpolarized light. (Note that if the excitation light is reflected off an aluminium surface surface rather than scattered into the detector, it will not be depolarized; this can sometimes lead to substantial errors [3]). However, the response function will differ significantly for horizontally or vertically polarized light [3], and since the emission from the sample will not in general be completely depolarized, the correction factors obtained for depolarized light will not be appropriate. The ratio of the transmission of the emission monochromator to vertically and horizontally polarized light is denoted $G(=T_V/T_H)$. When the fluorescence is viewed at 90° to the excitation beam, this can be measured experimentally. The emission of some suitable standard is measured through a polarizer oriented first vertically then horizontally. The factor G is then the ratio of these two intensities ($G=I_{HV}/I_{HH}$; this is independent of the emission anisotropy of the fluorophore used [7]).

G is found to vary very markedly with wavelength [33, 48], particularly with holographic grating monochromators [48]. It is well known that this must be taken into account in measurements of fluorescence polarization, but these instrumental effects may also influence normal fluorescence intensity measurements, depending on the emission anisotropy of the fluorophore.

7.4.2 Effects of emission anisotropy

The effect of emission anisotropy on the measured fluorescence intensity was noted by Weber and Teale in 1957 [16], and has periodically been discussed since [2, 9, 47, 49–53], though the general view is probably summarized by Chen [33]: 'the errors are usually neglected and indeed have never emerged as a great problem'.

Considering the common geometry in which the fluorescence is detected at 90° to the excitation beam, the upper limits of the error in intensity arising from emission anisotropy can be calculated from the basic equations given in [47]. The emission anisotropy, $r[= (I_\parallel - I_\perp)/(I_\parallel + 2I_\perp)]$ can have extreme values of 0.4 and -0.2; these would give measured intensities 0.90 and 1.05 times, respectively, that obtained for $r = 0$. Since one will be comparing a sample and a reference compound, errors of up to -20 per cent, + 10 per cent are possible from this source alone. Note that this calculation assumes unpolarized excitation and unbiased detection; in practice $G \neq 1$ and the errors can easily be twice as great as these.

The problems are somewhat different on the excitation side (quantum counters) and the emission side (emission standards). For a quantum counter, we require that not only the quantum yield but also the polarization is independent of excitation wavelength. This is very rarely the case; for example rhodamine B in dilute solution in glycerol shows polarization $[p = (I_\parallel - I_\perp)/(I_\parallel + I_\perp)]$ values of approximately -0.1 on excitation at 366 nm, -0.01 at 436 nm and +0.45 at 546 nm [33]. The substantial degree of polarization observed is due to the high viscosity of the solution which prevents the averaging out of the anisotropy by molecular tumbling during the lifetime of the excited state. In this respect it is unfortunate that the standard solvents for the rhodamine B quantum counter are glycerol or ethylene glycol. However, at the high concentrations employed substantial depolarization of emission by energy transfer would be expected, and the errors may be small. They do not appear to have been directly determined.

Emission standards present a different set of problems. In general, one would expect the polarization to be constant across an emission band, but it may vary markedly from one molecule to another. The fact that the emission is partially polarized will, however, bring into play the factor G, which is wavelength-dependent. (The purely instrumental effects can be removed by using quartz wedge depolarizers [54, 55] in the excitation and emission beams; the emission anisotropy error of -10 per cent, + 5 per cent noted above will, however, remain.) To obtain isotropic emission, one requires a small molecule with long excited state lifetime in a non-viscous solvent; for example, 10^{-4} M phenanthrene in methylcyclohexane gives completely unpolarized emission [56]; as noted above, this compound is not a suitable standard for general use). Clearly the use of rigid plastic or glass standards will potentially present serious polarization problems.

It has been proposed [50, 51] that viewing the fluorescence at an angle of 54.75° or 125.25° to the excitation beam removes all the effects of emission anisotropy, but this has been shown [47, 49] to be correct only for the case of unpolarized excitation and an unbiased detector. This and other [47] schemes involving detection of the emission at angles other than 90° to the excitation beam are difficult to implement in conventional fluorescence spectrometers. There remain three methods by which emission anistropy can be taken into account:

(a) Using polarizers in excitation and emission beams, measure the four components, I_H^H, I_V^H, I_H^V, I_V^V (where the superscript indicates the polarization on the excitation side, and the subscript that on the emission side). From these measurements one can obtain the factor G (see above), the degree of polarization of the excitation beam and the emission anistropy, and thus the *true* emission can be calculated.

(b) Using vertically polarized excitation, measure the emission through a polarizer oriented at 54.74° or 125.25° to the vertical [47, 57]; this measurement is independent of emission anisotropy.

(c) Using unpolarized excitation (obtained with a scrambler plate between monochromator and sample), measure the emission through a polarizer oriented at 35.25° to the vertical [57]; this measurement is independent of emission anisotropy.

On grounds of intensity, procedure (c) is perhaps to be preferred. It is of course absolutely essential in all these procedures to take account of the wavelength-dependence of the transmission of any polarizers, quartz wedges or scrambler plates used [48].

7.5 Recommendations

(a) The excitation source should be corrected by means of a quantum counter. Continuous correction, which also compensates for fluctuations in the light source, is particularly desirable, provided proper corrections are made for any beam-splitters used. The continuous correction system should be checked periodically with a quantum counter in the sample position.

(b) The two compounds which appear to be widely accepted as

suitable for quantum counter application are rhodamine B and 1-dimethylaminonaphthalene-5-sulphate. Measurements of their emission anisotropy are urgently required. Other quantum counters, particularly with an extended red response, are needed; rhodamine 101 may be preferable to rhodamine B.

(c) The primary calibration of the detector system is most conveniently done using the calibrated excitation system and a scatterer such as $BaSO_4$.

(d) There is also a need for compounds to serve as (secondary) emission standards. The only widely studied compound, quinine sulphate, has been the subject of much disagreement. Neither this nor any other compound can at present be said to have absolute emission characteristics sufficiently precisely defined by several laboratories to be recommended as standards. There is a considerable need for a process of certification of standard compounds. Among those which should be considered are tryptophan (emission 355 nm), p-terphenyl (340 nm), PBD (360 nm), tetraphenylbutadiene (460 nm) and 3-aminophthalimide (510 nm); again, compounds covering the red end of the spectrum are needed.

(e) The usefulness of standard compounds would be markedly increased if they were available as solid blocks. Both aromatic fluorophores in plastic and heavy metal doped inorganic glasses deserve further attention. The latter in particular cover the spectrum from 270-600 nm with plenty of overlap, and if their stability is satisfactory and the anisotropy problems are not too serious they should prove to be ideal standards. An alternative approach, which should also be pursued, is the use of solution standards sealed in cuvettes under nitrogen.

References

1 Parker, C. A. (1968) *Photoluminescence of solutions*, Elsevier, New York.
2 Demas, J. N., and Crosby, G. A. (1971) *J. Phys. Chem.*, **75**, 991.
3 Melhuish, W. H. (1972) *J. Res. Nat. Bur. Stand.*, **76A**, 547.
4 Rosen, P., and Edelman, G.M. (1968) *Rev. Sci. Instr.*, **36**, 809.
5 White, C. E., Ho, M., and Weimer, E. (1960) *Anal. Chem.*, **32**, 438.
6 Christensen, R. L., and Ames, I. (1961) *J. Opt. Soc. Am.*, **51**, 224.
7 Landa, I., and Kremen, J. C. (1974) *Anal. Chem.*, **46**, 1694.
8 Vavilov, S. I. (1927) *Z. Phys.*, **42**, 311.
9 Bowen, E. J. (1936) *Proc. Roy. Soc.*, Lond., **154A**, 349.
10 Melhuish, W. H. (1955) *New Zealand J. Sci. Tech.*, **37**, 142.
11 Melhuish, W. H. (1962) *J. Opt. Soc. Am.*, **52**, 1256.

12 Gill, J. E. (1969) *Photchem. Photobiol.*, **9**, 313.
13 Yguerabide, J. (1968) *Rev. Sci. Instr.*, **39**, 1048.
14 Demas, J. N. (1970) Ph.D. thesis, quoted in [2].
15 Taylor, D. G., and Demas, J. N. (1979) *Anal. Chem.*, **51**, 712.
16 Weber, G., and Teale, F. W. J. (1957) *Trans. Farad. Soc.*, **53**, 646.
17 Chen, R. F. (1967) *Anal. Biochem.*, **20**, 339.
18 Drexhage, K. (1977) in *Standardisation in Spectrophotometry and Luminescence Measurements*, ed. Mielenz, K. D., Velapoldi, R. A., and Mavrodineanu, R., Natl. Bur. Stand. Spec. Pub. No. 466, pp.33-40.
19 Taylor, D. G., and Demas, J. N. (1979) *Anal. Chem.*, **51**, 717.
20 Parker, C. A. (1958) *Nature*, **182**, 1002.
21 Cehelnik, E. D., and Mielenz, K. D. (1976) *Appl. Opt.*, **15**, 2259.
22 Hatchard, C. G., and Parker, C. A. (1956) *Proc. Roy. Soc. Lond.*, **235A**, 518.
23 Argauer, R. J., and White, C. E. (1964) *Anal. Chem.* **36**, 368.
24 Drushel, H. V., Sommers, A. L., and Cox, R. C. (1963) *Anal. Chem.*, **35**, 2166.
25 Mavrodineanu, R. (1977) in *Standardisation in Spectrophotometry and Luminescence Measurements*, ed. Mielenz, K. D., Velapoldi, R. A., and Mavrodineanu, R., Natl. Bur. Spec. Pub. No. 466, pp. 127-132.
26 Parker, C. A., and Hatchard, C. G., (1976) *Admiralty Materials Laboratory Report* AML-R7602.
27 Stair, R., Schneider, W. E., and Jackson, J. K. (1963) *Appl. Opt.*, **2**, 1151.
28 Parker, C. A., (1962) *Anal. Chem.*, **34**, 502.
29 Frei, R. W. (1977) in *Standardisation in Spectrophotometry and Luminescence Measurements*, ed. Mielenz, K. D. Velapoldi, R. A., and Mavrodineanu, R., Natl. Bur. Stand. Pub. No. 466, p.41.
30 Budde, W. (1977) in *Standardisation in Spectrophotometry and Luminescence Measurements*, ed. Mielenz, K. D., Velapoldi, R. A. and Mavrodineanu, R., Natl. Bur. Stand. Spec. Pub. No. 466, p. 75.
31 Eckerle, K. L., Venable, W. H., and Weidner, V. R. (1976) *Appl. Opt.*, **15**, 703.
32 Borreson, H. C., and Parker, C. A. (1966) *Anal. Chem.*, **38**, 1073.
33 Chen, R. F. (1972) *J. Res. Nat. Bur. Stand.*, **76A**, 593.
34 Velapoldi, R. A. (1972) *J. Res. Nat. Bur. Stand.*, **76A**, 641.
35 Birks, J. B. (1977) in *Standardisation in Spectrophotometry and Luminescence Measurements*, ed. Mielenz, K. B., Velapoldi, R. A., and Mavrodineanu, R., Natl. Bur. Stand. Spec. Pub. No. 466, pp. 1-11.
36 Velapoldi, R. A. and Mielenz, K. D. (1978) *Abstract no. 102, 29th Pittsburgh Conference*, Cleveland, USA.
37 West, M. A. and Kemp, D. R. (1977) *Amer. Lab.*, **9**, 37.
38 Lippert, E., Nägele, W., Siebold-Blankenstein, I., Staiger, U., and Voss, W. (1959) *Z. Analyt. Chem.*, **170**, 1.

39 Berlman, I. B. (1971) *Handbook of Fluorescence Spectra of Aromatic Molecules.* 2nd edn. Academic Press, New York.
40 Rusakowicz, R. and Testa, A. C. (1968) *J. Phys. Chem.*, **72**, 2680.
41 Parker, C. A. and Rees, W. T. (1960) *Analyst*, **85**, 587.
42 Melhuish, W. H. (1961) *J. Phys. Chem.*, **65**, 229.
43 Ware, W. (1976) *Chem. Phys. Lett.*, **39**, 449.
44 Williams, D. L. and Heller, A. (1970) *J. Phys. Chem.*, **74**, 4473.
45 Melhuish, W. H. (1964) *J. Opt. Soc. Am.*, **54**, 183.
46 Reisfeld, R. (1972) *J. Res. Nat. Bur. Stand.*, **76A**, 613.
47 Mielenz, K. B., Cehelnik, E. D., and McKenzie, R. L. (1976) *J. Chem. Phys.*, **64**, 370.
48 Dale, R. E. (1979) personal communication.
49 Cehelnik, E. D., Mielenz, K. D., and Velapoldi, R. A. (1975) *J. Res. Nat. Bur. Stand.*, **79A**, 1.
50 Almgren, M. (1968) *Photochem. Photobiol.*, **8**, 231.
51 Shinitzky, M. (1972) *J. Chem. Phys.*, **56**, 5979.
52 Paoletti, J., and Le Pecq, J-B. (1969) *Anal. Biochem.*, **31**, 33.
53 Kalantar, A. H. (1968) *J. Chem. Phys.*, **48**, 4992.
54 Rahn, L. A., Temple, P. A., and Hathaway, C. E. (1971) *Appl. Spectrosc.*, **25**, 675.
55 Reed, P. R., and Landon, D. O. (1972) *Appl. Spectrosc.*, **26**, 489.
56 Azumi, T., and McGlynn, S. P. (1962) *J. Chem. Phys.*, **37**, 2413.
57 Spencer, R. D., and Weber, G. (1970) *J. Chem. Phys.*, **52**, 1654.

8 The determination of quantum yields

8.1 Introduction

Luminescence quantum yield is defined as the fraction of molecules that emit a photon after direct excitation by a light source. This quantity is not strictly identical to the total number of emitted photons which escape a bulk sample divided by the total number of absorbed photons although in many instances the two quantities are practically equal [1]. Both theoretically and practically quantum yield values are useful. They provide information on radiationless processes in molecules, and, in the assignment of electronic transitions, are of use in assessing the potential of the fluorimetric determination of materials, assessing their purity and for judging the suitability of materials as wavelength shifters.

The measurement of the quantum yield of a compound in solution is affected by a number of factors. It is important to understand these in order to appreciate the advantages and pitfalls of individual methods of determining quantum yields and the relative merits of individual fluorescence standards. Many of the factors have been treated in detail in earlier chapters.

8.1.1 *Inner filter and self-absorption effects*

The linear relationship between the concentration of a solute and its fluorescence intensity only applies in solutions of very low absorbance. At higher concentrations, correction factors, dependent upon the geometry of the instrument cell compartment, must be applied. In addition, if there is a considerable overlap between the excitation and fluorescence spectra, self-absorption may occur; again this effect is negligible only in very dilute solution. Several materials that have been proposed as fluorescence standards (Chapter 7) show self-absorption effects. Inner filter and self-absorption phenomena have been treated in detail in Chapter 5.

8.1.2 Exciting light wavelength effects

It is commonly assumed that fluorescence quantum yield of a compound will be independent of the excitation wavelength. This will be the case if the transitions from the higher excited states S_2, S_3, S_4 etc. to the S_1 (first excited state) have a quantum yield of 1 (Vavilov's Law). A number of deviations from this situation have been observed for solutions of benzene, 1,6-dimethylnaphthalene, fluorobenzene, 2-methylnaphthalene, mesitylene, naphthalene, phenylalanine, tryptophan and tyrosine [2]. These exceptions to Vavilov's Law emphasize the importance of verifying its applicability for each proposed standard. If the quantum yield is found to be dependent on the excitation wavelength, then only that part of the excitation spectrum where $S_0 \to S_1$ transitions occur must be used (this will also tend to lessen photodecomposition effects, see Chapter 6). Large errors can be introduced into the determination of quantum yields if the exciting light is non-monochromatic. For example, the mercury line at 313 nm is weakly absorbed by anthracene but the absorption of the mercury line at 254 nm is over 100 fold greater. Thus if the 313 nm line is isolated by filters that transmit one percent of the 254 nm line, the observed intensity will be double that which would be observed if a pure 313 nm line were used for excitation (cf. Section 3.1). Such effects can also have a profound influence on the comparison of quantum yields since it is unlikely that the substances for comparison will have the same absorption profile. A similar problem is found using xenon light sources whose wavelength bands are isolated by monochromators for the band passes illuminating the cuvette are often quite broad. A good monochromatic light source is needed especially for compounds which have a very spiky excitation spectrum. A broad excitation spectrum with minimal vibrational bands is ideal.

8.1.3 Refractive index

A correction for differences in refractive index between fluorescence standard and sample may be required in order to compensate for differences in solution optical geometry. The errors due to refractive index differences are two-fold: (a) as radiation passes from a high to a low refractive index (i.e. from solution to air) a change in intensity occurs at the interface, because light is bent away from the normal; (b) internal reflection of emitted light within the cuvette may occur at all the interfaces between zones of different refractive indices [2]. Such errors may occur if different solvents are used for standards and samples. It has been suggested that if the fluorescence of the standard and sample are measured at 55° to the angle of the exciting light, refractive index differences will be minimized [2].

Melhuish [3] has demonstrated that by using a cuvette with the back and side painted with a flat black paint, much of the internal reflection error can be eliminated. A correction factor may be applied for the effects of emitted light traversing different refractive indices with a 90° viewing angle i.e.

$$(I)_o = \frac{(I)_i \cos\theta}{n_i(n_i^2 - \sin^2\theta)^{1/2}}$$

where n_i = refractive index of medium. The observed intensity in air $(I)_o$ is related to the intensity in the cuvette $(I)_i$; θ is the angle of deviation of light from the normal [4].

8.1.4 Polarization/anisotropy effects

When a fluorescence system is excited by unpolarized light and its emission is measured without a polarizer, the emission intensity has a typical anisotropic distribution which is directly related to its degree of polarization. This effect can introduce an error of up to 20 per cent in quantum yield measurements [5], and has been described in detail in Chapter 7 (Section 7.4).

8.1.5 Temperature effects

It is important to be aware of the (normally negative) temperature coefficient of fluorescence intensity of the compounds under study. Such effects can be appreciable (see Chapter 6, Section 6.1).

8.1.6 Sample/solvent impurities

High purity materials are normally required in preparing solutions for quantum yield determinations. In cases where quantum yields are low, particular care must be taken to avoid problems arising from scattered light, for example, from dust contamination of the samples. Traces of absorbing impurities may cause particular problems in spectral regions where the sample has low absorption, for example, fluorescein in 0.1 M NaOH has a very low absorption in the 340–370 nm region [4]. Fluorescent blanks can arise from sample, solvent or cuvette. Blanks due to solvent and cuvette can be detected readily and should always be checked. If their contribution to the fluorescence signal is significant and cannot be readily overcome, the blank can be deducted from the fluorescence signal of the sample before spectral correction is attempted. Fluorescent contaminants in the compounds used to make up solutions of sample or standard are not so readily detected. Quenching may also go undetected. With substances with a relatively long-lived fluorescence, quenching by dissolved oxygen may be appreciable, for example, many aromatic hydrocarbons. Oxygen quenching may be overcome by passing a current of pure nitrogen through the sample prior to measurement (see Chapter 5). Quenching by other contaminants may also cause problems, for example, chloride ions in the sulphuric acid used to prepare quinine sulphate standards.

8.1.7 Photodecomposition and chemical stability

Photochemical stability tends to decrease with increasing sample dilution and lowering of the excitation wavelength (see Chapter 8, Section 8.2). Solutions placed in cuvettes in which the bulk of the solution is irradiated simultaneously (for example, microcuvettes) will show apparently greater photodecompostion than those in which only a small part of the solution is irradiated at any one time. Many fluorescence standards photodecompose e.g. quinine sulphate, tryptophan.

Many of the xanthene dye compounds, e.g. eosin, tend to decompose in alkaline solution. Storage of dilute solutions of standards in organic solvents may result in appreciable adsorption of compound on to the walls of the storage vessel. Storage vessels of soft glass may be particularly bad in this respect. A number of the above problems can be minimized by using a relatively concentrated stock solution of standard which is diluted appropriately just before use.

8.2 Primary methods of determining quantum yields

This discussion will be confined to determining the quantum yields of substances in solution. For solid samples, the reader is referred to a number of reviews [1, 6]. The history of quantum yield determinations shows clearly that it is very difficult to eliminate all errors and that no ideal method of determination exists. A number of independent methods are therefore required in order to verify quantum yield values for each compound.

8.2.1 Light scattering methods

Ideal scatterers behave as substances which re-emit all photons with which they are irradiated without a change in wavelength, that is, for practical purposes they can be considered to have a quantum efficiency of unity.

Two scattering sources have been studied in particular, namely: magnesium oxide-coated glass and colloidal solutions of glycogen or silica.

Magnesium oxide coated glass scatterer was first proposed as a reference material to calibrate the quantum yield of compounds in solution by Vavilov in 1924 [7]. Magnesium oxide scatterers suffer from a number of disadvantages [1] as primary reference materials:

(a) magnesium oxide coatings tend to age rapidly,
(b) the reflectance is dependent on the method of preparation and thickness of the coating,
(c) the reflectance falls off markedly in the UV region.

In addition, problems can arise from comparing light reflected from a solid surface with light emission from a fluorescent compound in solution in a cuvette; of particular note are polarization variations between standard and sample and internal reflection of light within the cuvette (but not the standard). Errors of up to 20 per cent may occur as a result of reflection differences [1]. To avoid the problems involved in calibrating the monochromator, Melhuish [8] used a rhodamine B quantum counter as the detector system. Although corrections can be introduced for many of these variables, they are tedious to apply and except for optically dense solutions colloidal solutions are probably better primary reference standards.

Colloidal solution scatterers were first introduced by Weber and Teale [9] in 1958. A particular appeal of this technique is that in principle, many commercial fluorimeters can be used without modification. Essential correction factors in comparing the signal from the scatterer standard and the fluorescent sample are:

(a) a correction for the wavelength sensitivity of the detector system (normally unnecessary with a quantum counter detector)
(b) a correction for polarization differences between standard and sample (which is more or less constant for many fluorescent solutions as they freely rotate between being excited and emitting light)

(c) differences in refractive indices and optical densities between standard and sample.

In selecting the light scattering standard consideration must be given to:

(a) possible light absorbing and fluorescing contaminants. (These can be tested for by applying Rayleigh's Law i.e. absorbance x λ^4 should be a constant)

(b) the most appropriate means of determining the absorbance of the scatterer standard and the fluorescent sample. (Fluorescent samples with high quantum efficiencies will tend to produce absorbance readings below the true values while accurate measurement of absorbance on highly scattering samples may require careful choice of spectrophotometer).

(c) stability of the scatterer solutions e.g. both glycogen and Ludox tend to hydrolyse on storage.

Dawson and Kropp [10] have attempted to overcome the absorbance problem by plotting detector reading of sample against its absorbance. This approach gives a much clearer picture of errors from reabsorption etc.

Pure horse muscle and cat liver glycogen or colloidal silica (Ludox, Du Pont) are the most widely used scatterer solutions. At wavelengths below 300 nm both materials deviate from ideal scatter behaviour and are not recommended [7, 11, 12]. The rationalization of the many criticisms of the Weber and Teale method lies in its use by many workers at wavelengths below 300 nm.

8.2.2 Absolute evaluation of the geometry

This approach is valuable for optically dense solutions but less satisfactory for dilute solutions. Instead of using a light scatterer (Section 8.2.1), a quantum counter/detector is introduced in the cuvette position to measure the intensity of exciting light directly [13]. This method avoids the problem of determining the scatterer's reflectance properties. Although the method is very demanding on the quantum counter/detector, suitable arrangements can be obtained commercially. An adaptation of this approach is to use a chemical actinometer.

The principle of this method is that the intensity of light falling in the sample cuvette is determined using a chemical actinometer. The sample is then placed in position and irradiated while it is almost entirely surrounded by the same chemical actinometer thereby enabling the amount of emitted light to be detected. The technique could be applied to several commercial fluorescence spectrometers or spectrophotometers but considerable adaptation of the sample compartment would be required.

In the system developed by Demas and Blumenthal [14], a laser light source was employed. A small correction factor was introduced to allow for light emitted in the small area at the front and back of the cuvette not covered by the actinometer.

Of considerable importance are the methods using an integrating sphere [1] as part of the detection system: the combination of integrating sphere, monochromator and photodetector is calibrated using a standard light source. Again, some difficulty may be found in adapting all but the latest commercially-available instruments to this approach.

8.2.3 Calorimetric methods

Calorimetric measurements [1, 15] are very largely independent of the errors inherent in optical techniques and therefore most usefully complement other means of determining quantum yields. The principle of the calorimetric approach is that temperature (or volume) changes are compared during irradiation of a non-luminescent standard and the fluorescent sample of very similar optical densities. As the non-luminescent standard will have zero quantum efficiency, the ratio of the temperature (or volume) changes of the two solutions gives the fraction of the absorbed energy which is lost by non-radiative processes in the fluorescent sample (that is, the complement of the quantum yield). The major difficulties with calorimetric methods lie in:

(a) The relative insensitivity of the detectors (although these are fast improving) which necessitates the use of relatively concentrated solutions.

(b) Selection of a suitable light source. A non-continuous intense stable lamp with well separated lines and minimal infra-red radiation is needed; laser sources are probably most suitable.

(c) A specialized sample chamber is required to eliminate ambient temperature effects.

8.3 Secondary methods of determining quantum yields: use of fluorescence standards

Rather than the methods described in Section 8.2, many workers prefer to use comparative methods of determining quantum yields. Such methods are based on the fact that if two substances 1 and 2 are studied in the same apparatus, and using the same incident light intensity, the integrated areas under their corrected fluorescence spectra (S_1 and S_2) are simply related as follows:

$$\frac{S_2}{S_1} = \frac{\phi_2}{\phi_1} \frac{A_2}{A_1}$$

where ϕ values are quantum yields and A values absorbances at the respective excitation wavelengths. The application of this method, including the further corrections needed for refractive index and polarization effects, has been thoroughly discussed elsewhere [1, 2, 16].

As corrected spectra (Chapter 7, Section 7.3) are becoming more readily available, this method is likely to remain the most popular approach to quantum yield determinations. However, it requires a series of suitable standard materials if it is to be used over a range of wavelengths. The ideal properties of standards have been listed in Section 7.3.2. Most, though not all, of the criteria may be met by dilute solutions of appropriate compounds, but some of the requirements are mutually exclusive, and no compound has been found to be ideal. For example, materials with a large Stokes shift may have low quantum yields, while materials with long fluorescence lifetimes, while exhibiting minimal polarization effects, are more subject to quenching interactions.

Many of the commonly-used standards have been discussed in the previous chapter, and some spectra are presented in the Appendix. Quinine sulphate is still probably the popular standard material, but it has a number of disadvantages, most notably photochemical instability, a fluorescence intensity depending on sulphuric acid concentration and an emission spectrum that depends on the excitation wavelength. Demas and Crosby [1] have concluded that the best value of the quinine sulphate quantum yield is 0.546 (excitation wavelength 365 nm, solvent 1 N sulphuric acid, concentration $<10^{-4}$M, temperature 25°C). Two other compounds have found wide use as fluorescence standards, namely: fluorescein and diphenylanthracene. Fluorescein (ϕ_f = 0.91, [1]) suffers from serious self-absorption problems because of the strong overlap of the excitation and fluorescence spectra: in addition, it is not very stable and has low extinction coefficients in the useful 350 nm region. Diphenylanthracene (DPA) has recently become very popular. Morris et al. [17] have found its quantum yield to be 0.95 in ethanol. The solvent has to be deoxygenated but, as demonstrated in Chapter 5 (Section 5.2) this presents no great difficulty. DPA is relatively easily purified [18] and although it is susceptible to self-absorption, advantage can be taken of the high extinction coefficient, and very dilute solutions can be used.

Other possible quantum yield standards not listed in Chapter 7 include 4-dimethylamino-4'-nitrostilbene in o-dichlorobenzene [19], and perylene. The former compound has the advantage of fluorescing at wavelengths above 700 nm, a region not covered by other standards. Perylene has a high quantum yield (0.92 [1]) and the advantages and disadvantages of other hydrocarbon standards: this material and a number of others, notably the conventional organic scintillants, deserve further study as possible standards. Other suggested materials have included benzophenone [20], naphthalene [21], acridone [2], 9-amino-acridine [2], 2-aminopurine [22], methylene blue [23], chromium complexes [24], 7-hydroxycoumarin and its 4-methyl derivative [25] and terbium and europium chelates [26].

8.4 Other methods of determining quantum yields

In addition to the methods described above, at least two other methods have been described recently for quantum yield studies. Britten et al. [27] developed an ingenious method which avoids direct measurement of the absorbance of the sample. Instead, the fluorescence intensity is measured at two points along the excitation path, and the ratio of the two fluorescence intensities related to the sample absorbance (cf. Chapter 5, Section 5.1). Some modification to the sample compartment of the fluorescence spectrometer is required, but the method is nonetheless convenient, and yields values in reasonable agreement with other approaches. A related method, requiring no instrumental modifications, has been studied by Gains and Dawson [28].

A completely different approach, that of time-correlated single photon counting, has been suggested by Upton and Cline Love [29]. A pulsed source is needed, and the same instrument is used for absorbance and fluorescence measurements. Again the agreement with results obtained by classical techniques is good. The equipment needed, however, it not available in many laboratories.

8.5 Summary and recommendations

(a) Accurate quantum yields of species in solution are of value in many photochemical, biochemical and analytical studies, and should be available for a wide range of materials.

(b) None of the methods currently available is ideal, but the comparative technique (Section 8.3) will remain the most frequently used.

(c) Considerable further efforts are needed to identify new quantum yield secondary standards, and to characterize further those that are already in use.

(d) Of the secondary standards in use at present, quinine sulphate may have too many disadvantages, and diphenylanthracene may be preferable.

References

1. Demas, J. N. and Crosby, G. A. (1971) *J. Phys. Chem.*, **75**, 991.
2. Birks, J. B. (1977) in *Standardisation in Spectrophotometry and Luminescence Measurements* (ed. K. D. Mielenz, R. A. Velapoldi, and R. Mavrodineanu) U.S. Dept. of Commerce p.1.
3. Melhuish, W. H. (1961) *J. Opt. Soc. Amer.*, **51**, 278.
4. Parker, C. A. (1968) *Photoluminescence of Solutions,* Elsevier, New York, p. 261.
5. Shinitzky, M. J. (1972) *J. Chem. Phys.*, **56**, 5979.
6. Bril A. and de Jager-Veenis, A. W. (1977) in *Standardisation in Spectrophotometry and Luminescence Measurements* (ed. K. D. Mielenz, R. A. Velopoldi, and R. Mavrodineanu) U.S. Dept. of Commerce p.13.
7. Vavilov, S. I. (1924) *Z. Phys.*, **22**, 266.
8. Melhuish, W. H. (1955) *New Zealand J. Sci. Tech.*, **37**, 142.
9. Weber, G. and Teale, F. W. J. (1957) *Trans. Farad. Soc.*, **53**, 646.
10. Dawson, W. R. and Kropp, J. L. (1965) *J. Opt. Soc. Amer.*, **55**, 822.
11. Dawson, W. R. and Windsor, M. W. (1968) *J. Phys. Chem.*, **72**, 3251.
12. Chen, R. F. (1967) *Anal. Lett.*, **1**, 35.
13. Andreeschev, E. A. and Rozman, I. M. (1960) *Opt. Spectrosc.*, **8**, 435.
14. Demas, J. N. and Blumenthal, B.H. (1977) in *Standardisation in Spectrophometry and Luminescence Measurements* (ed. K. D. Mielenz, R. A. Velapoldi, and R. Mavrodineanu) U.S. Dept. of Commerce p.21.
15. Callis B. J. (1977) in *Standardisation in Spectrophometry and Luminescence*

Measurements (ed. K. D. Mielenz, R. A. Velapoldi, and R. Mavrodineanu) U.S. Dept of Commerce p. 25.
16 Parker, C. A. and Rees, W. T. (1960) *Analyst (London)*, **85**, 587.
17 Morris, J. V., Mahaney, M. A. and Huber, J. R. (1976) *J. Phys. Chem.*, **80**, 969.
18 Lloyd, J. B. F. (1979) *J. Chromatog.*, **178**, 249.
19 White, C. E. and Argauer, R. J. (1970) *Fluorescence Analysis: A Practical Approach*, Marcel Dekker, New York.
20 Dubois, J. T. and Wilkinson, F. (1963) *J. Chem. Phys.*, **39**, 899.
21 Ermolaev, V. L. and Svitashev, K. K. (1959) *Opt. Spectrosc.*, **7**, 399.
22 Drobnik, J. and Yeargers, F. (1966) *J. Mol. Spectrosc.*, **19**, 412.
23 Seeby, G. R. (1969) *J. Phys. Chem.*, **73**, 125.
24 Chatterjee, K. K. and Forster, L. S. (1964) *Spectrochim. Acta*, **20**, 1603.
25 Dawson, W. R., Kropp, J. L. and Windsor, M. W. (1966) *J. Chem. Phys.*, **45**, 2410.
26 Petrovich, P. I. and Borisevich, N. A. (1963) *Bull. Acad. Sci. USSR Phys. Ser.*, **27**, 701.
27 Britten, A., Archer-Hall, J. and Lockwood, G. (1978) *Analyst (London)*, **103**, 928.
28 Gains, N. and Dawson, A. P. (1979) *Analyst (London)*, **104**, 481.
29 Upton, L. M. and Cline Love, L. J. (1979) *Anal. Chem.*, **51**, 1941.

Appendix Corrected excitation and emission spectra

This appendix includes the corrected excitation and fluorescence spectra of a number of the solutes listed in Chapters 7 and 8 as possible standards for quantum yield determinations and for spectral correction of other solutes.

The spectra were obtained using a Perkin-Elmer MPF 44B fluorescence spectrometer, further equipped with a DCSU-2 Differential Corrected Spectra Unit. The spectrometer is fitted with a 150 W xenon lamp operated as a continuous wave source, slits continuously variable to produce spectral bandwidths of 0.2 nm to 20 nm, and an R928 photomultiplier. The excitation and emission monochromators are Czerny-Turner mounted $f/4$ grating monochromators, ruled at 1 200 lines mm^{-1}, and blazed for maximum efficiencies at 280 nm and 350 nm, respectively. The photometric system can be used in 'direct' (energy) or ratio recording modes, the latter mode being used to record corrected spectra. In the present work the instrument was used in conjunction with a model 056 recorder. Spectra were recorded at a scale of 1 mm $\equiv$ 1 nm and traced directly for reproduction in this appendix.

When the spectrometer is used to obtain corrected spectra, the output of the xenon lamp is obtained using rhodamine B as a quantum counter (cf. Chapter 7, Section 7.2.2). This output curve is stored in the DCSU-2 unit and can subsequently be used to correct excitation spectra, the correction being applied at 0.2 nm intervals. Corrected emission spectra can be obtained either with the aid of a calibrated tungsten lamp, or by using the previously-calibrated xenon lamp, whose light is deflected into the emission monochromator by a diffuser plate mounted in the sample position. The latter procedure was used in the present work (cf. Section 7.3.1) but no corrections were made for polarization effects (Section 7.4).

80 *Standards in fluorescence spectrometry*

The spectra in Figs. A.5–A.17 were obtained using the following general procedures:

(a) All measurements were made at 25.0 ± 0.5°C.

(b) Samples and solvents were of the highest available purity. (Sources and melting points are given with each spectrum). Water was distilled three times from silica and filtered through a 0.22 μm membrane filter. Ethanol (96% UV grade; James Burroughs Ltd, London) was distilled three times before use.

(c) Wherever possible the absorbance of the samples was maintained below 0.05 units throughout the wavelength range studied (cf. Chapter 5). To obtain a satisfactory signal-to-noise ratio it was thus necessary to utilize different bandwidths on the spectrometer. The bandwidths used are given on each spectrum.

(d) The wavelength scales on the spectrometer were periodically checked using the procedures described in Section 2.5 (samples used were coronene, rhodamine B, and 7, 8-benzoquinoline in the form of polymer blocks) and Section 2.3, i.e. using the xenon lamp emission lines.

These precautions were found to be essential to the recording of reliable corrected spectra. Several instances of artefacts caused by impure samples and solvents, and excessive absorbances, were noted. Thus, ethanol used without repeated distillation yielded significant fluorescence signals; and a commercial sample of phenanthrene was found to be contaminated by anthracene. It must be appreciated that, by its very nature as an ultra-sensitive method, fluorimetry presents difficulties in connection with the purity of standard materials. Samples which appear to be pure by conventional criteria such as melting point, infra-red and n.m.r. spectra, elemental analysis, chromotographic homogeneity, etc., may nonetheless contain traces of fluorescent (or quenching) impurities. This important aspect of fluorescence standardization will be dealt with at length in a subsequent volume of this series.

The validity of the spectra obtained was tested in two ways. In all cases the corrected excitation spectrum was compared with the absorption spectrum of the compound (see, for example, Figs. A.1–A.3). Absorption spectra were obtained using a Pye-Unicam Model 8-100 spectrometer. The limitations of this comparison procedure are considerable (see Section 7.2.3). Nonetheless the correspondence between the spectra obtained suggests that the corrected excitation spectra reproduced here do not suffer from any substantial errors at wavelengths >250 nm. At very low wavelengths

Table A.1: *The corrected emission spectrum of quinine sulphate dihydrate in 0.105 M perchloric acid between 375 nm and 575 nm*

λ (nm)	$E(\lambda)$ [1]	$E(\lambda)$ This work	λ (nm)	$E(\lambda)$ [1]	$E(\lambda)$ This work
375	0.005	0.008	480	0.782	0.808
380	0.012	0.015	485	0.719	0.739
385	0.028	0.034	490	0.659	0.675
390	0.057	0.068	495	0.595	0.608
395	0.103	0.120	500	0.541	0.551
400	0.170	0.177	505	0.486	0.508
405	0.257	0.268	510	0.434	0.454
410	0.359	0.376	515	0.386	0.399
415	0.471	0.494	520	0.342	0.355
420	0.586	0.608	525	0.302	0.311
425	0.694	0.716	530	0.264	0.275
430	0.792	0.814	535	0.231	0.241
435	0.874	0.900	540	0.201	0.212
440	0.940	0.957	545	0.175	0.182
445	0.984	0.988	550	0.153	0.161
450	0.999	0.999	555	0.132	0.136
455	0.997	0.999	560	0.116	0.118
460	0.982	0.982	565	0.101	0.105
465	0.947	0.959	570	0.080	0.090
470	0.897	0.924	575	0.076	0.085
475	0.838	0.858			

agreement is less good, possibly because of high stray-light levels in both instruments. The corrected emission spectrum of quinine sulphate was compared with the data very recently published by Velapoldi and Mielenz [1] in a comprehensive monograph (Fig. A.4). The tabulated comparison (Table A.1) shows that the differences between the two sets of data did not exceed approximately 5% over the wavelength range 400—550 nm, and did not exceed approximately 3% between 425 nm and 500 nm. Again the acceptable accuracy of the spectra in this Appendix is implied.

These spectra are thus probably typical of the corrected spectra obtainable with a modern commercially-available instrument. The wavelengths of the excitation and emission maxima have an estimated accuracy of ± 2 nm.

Reference

1 Velapoldi, R. A. and Mielenz, K. D. (1980), *NBS Special Publication* number 260-64, US Department of Commerce.

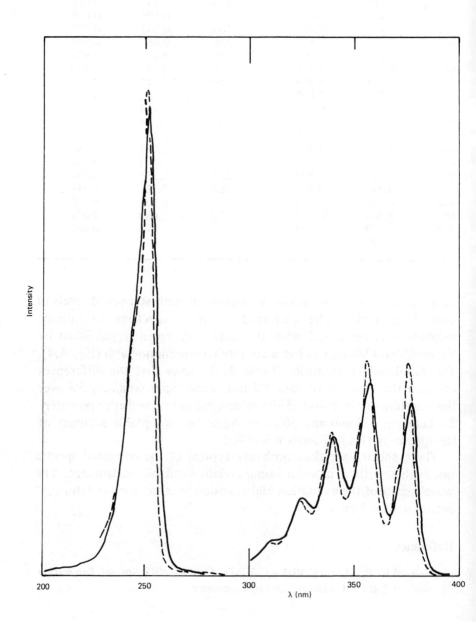

Fig. A.1 Comparison of the corrected excitation spectrum (———) and the absorption spectrum (- - - -) of anthracene in tri-distilled ethanol. For the excitation spectrum the instrument gain was 0.3, the fluorescence wavelength 404 nm and the monochromator bandwidths 4 nm (excitation) and 12 nm (emission). Sample concentrations were: (i) 200–300 nm: 2.85×10^{-7} M (maximum absorbance 0.045 at 252 nm) and (ii) 300–400 nm: 5.7×10^{-6} M (maximum absorbance, 0.040 at 356 nm). For the absorption spectrum the full scale deflection corresponded to 1.0 absorbance unit: concentrations were 5.7×10^{-6} M (200–300 nm) and 5.7×10^{-5} M (300–400 nm).

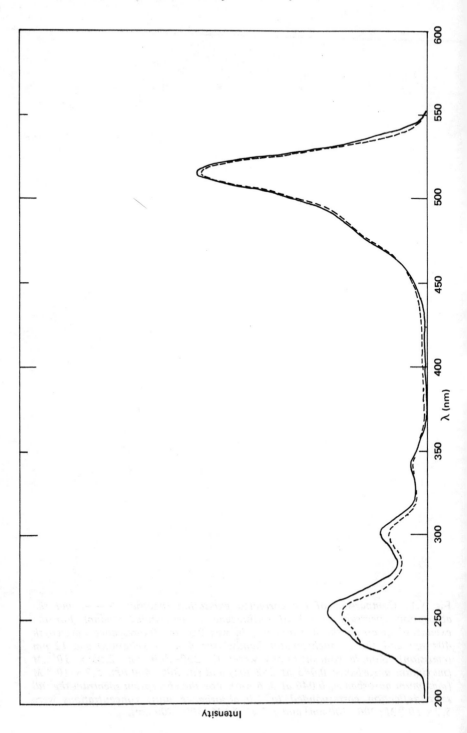

Fig. A.2 *Comparison of the corrected exciation spectrum (———) and the absorption spectrum (- - - -) of eosin in 0.1 M aqueous NaOH. The excitation spectrum was obtained as in Fig. A.8, and the absorption spectrum was obtained using a $1.96 \times 10^{-5} M$ solution; full scale deflection in the latter spectrum corresponds to 2.0 absorbance units.*

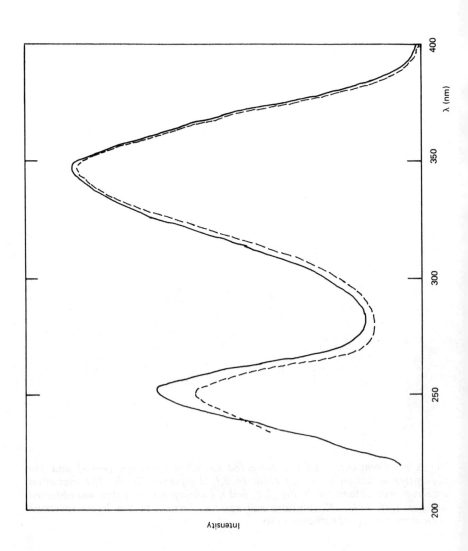

Fig. A.3 Corrected excitation (———) and absorption (- - - -) spectra of 1,1,4,4-tetraphenylbutadiene. The excitation spectrum was obtained as in Fig. A.15. The absorption spectrum was taken using a 3×10^{-5} M solution in cyclohexane at a scan rate of 1 nm s^{-1}.

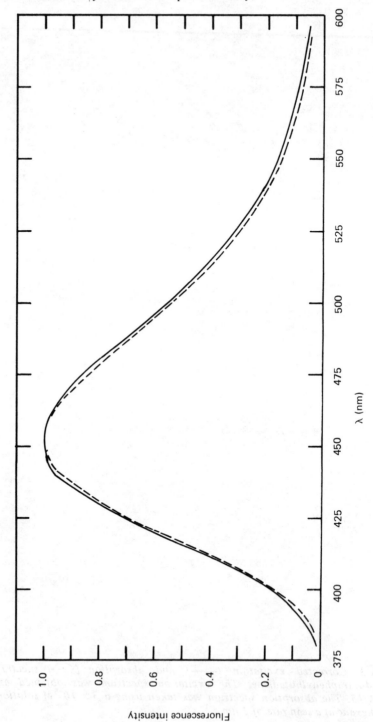

Fig. A.4 *Comparison of the corrected fluorescence spectra of quinine bisulphate in 0.105 M perchloric acid obtained in the present work (———), and in [1] (- - - -). Each spectrum is normalized to a value of 1.000 at the wavelength of maximum fluorescence.*

90 *Standards in fluorescence spectrometry*

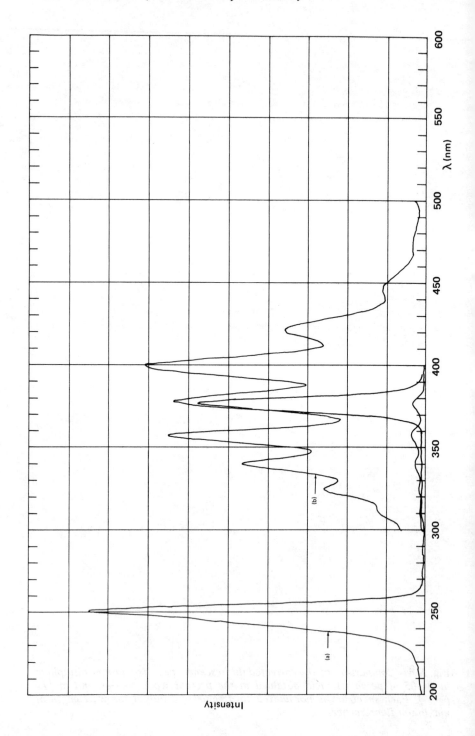

Fig. A.5 ANTHRACENE

Source: British Drug Houses Ltd
Melting point: 214–217°C
Solvent: Tri-distilled ethanol
Concentration: 2.85×10^{-7} M
Excitation spectrum: λ_f = 404 nm
    Bandwidths: 4 nm (excitation monochromator)
                     12 nm (emission monochromator)
    Gain: 0.3 (a), 3 (b) (see also Fig. A.1)
Emission spectrum: λ_{ex} = 252 nm.
    Bandwidths: 4 nm (excitation monochromator)
                     8 nm (emission monochromator)
    Gain: 0.3
    Maximum absorbance: 0.045 (252 nm)

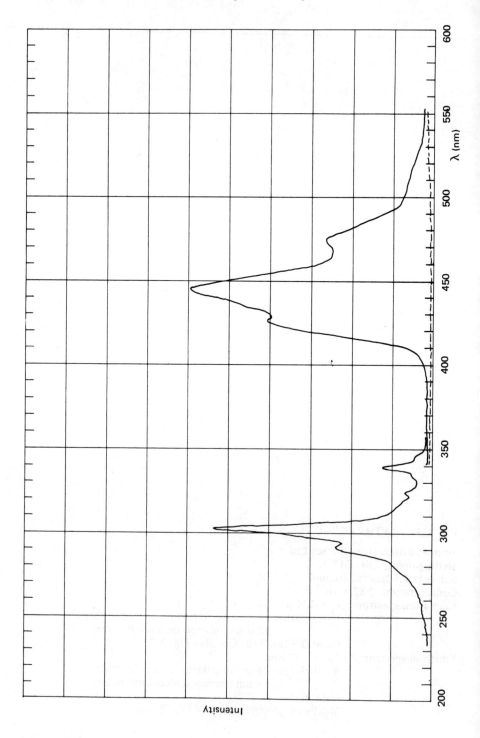

Fig. A.6 CORONENE

Source: Koch-Light Laboratories Ltd
Melting point: 360° (decomp)
Solvent: Tri-distilled ethanol (broken line)
Concentration: 2.4×10^{-7} M
Maximum absorbance: 0.05 (302 nm)
Excitation spectrum: λ_f = 450 nm
                            Bandwidths: 2 nm (excitation monochromator)
                                       6 nm (emission monochromator)
                            Gain: 3
Emission spectrum: λ_{ex} = 302 nm
                            Bandwidths: 2 nm (excitation monochromator)
                                       16 nm (emission monochromator)
                            Gain: 1

94 *Standards in fluorescence spectrometry*

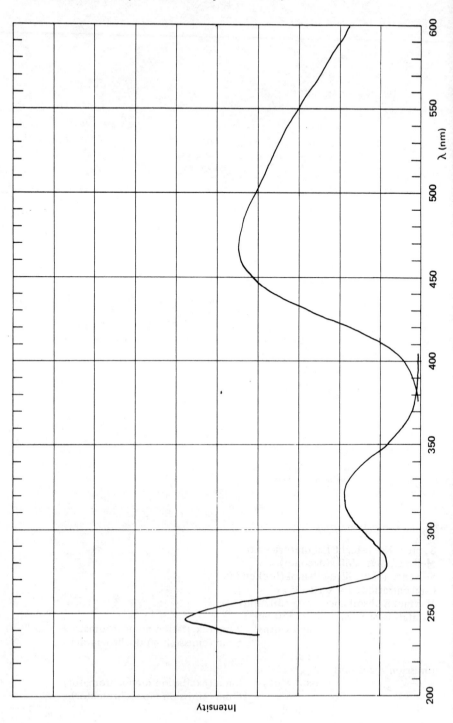

Fig. A.7 *DIMETHYLAMINONAPHTHALENE SULPHONYL CHLORIDE*

Source: British Drug Houses Ltd
Melting point: 69-71°C
Solvent: Tri-distilled ethanol
Concentration: 1.48×10^{-6} M
Excitation spectrum: λ_f = 455 nm
                      Bandwidths: 4 nm (excitation monochromator)
                                     12 nm (emission monochromator)
                      Gain: 3
Emission spectrum: λ_{ex} = 320 nm
                      Bandwidths: 4 nm (excitation monochromator)
                                     12 nm (emission monochromator)
                      Gain: 1

96 Standards in fluorescence spectrometry

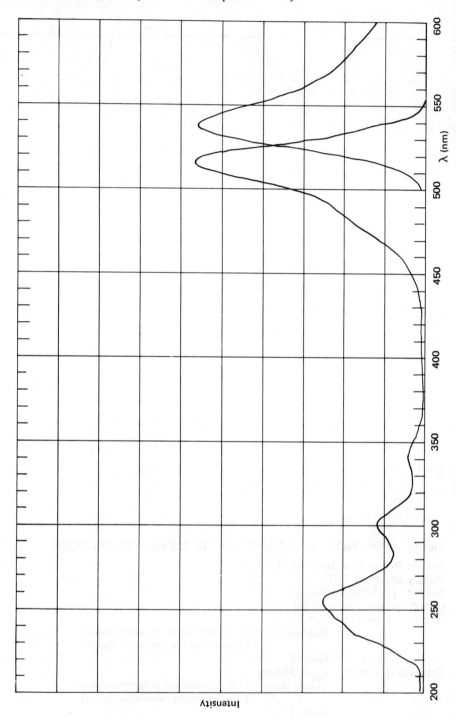

Fig. A.8 *EOSIN*

Source: British Drug Houses Ltd, (CI 45380)
Melting point: 300°C (decomp)
Solvent: 0.1 M aqueous NaOH
Concentration: 7.84 x 10^{-7}M
Excitation spectrum: λ_f = 535 nm
        Bandwidths: 4 nm (excitation monochromator)
              6 nm (emission monochromator)
        Gain: 1
        Maximum absorbance: 0.044 (515 nm)
Emission spectrum: λ_{ex} = 255 nm
        Bandwidths: 4 nm (excitation monochromator)
              6 nm (emission monochromator)
        (Cut-off filter employed to eliminate second order
        scattering signal)

98 *Standards in fluorescence spectrometry*

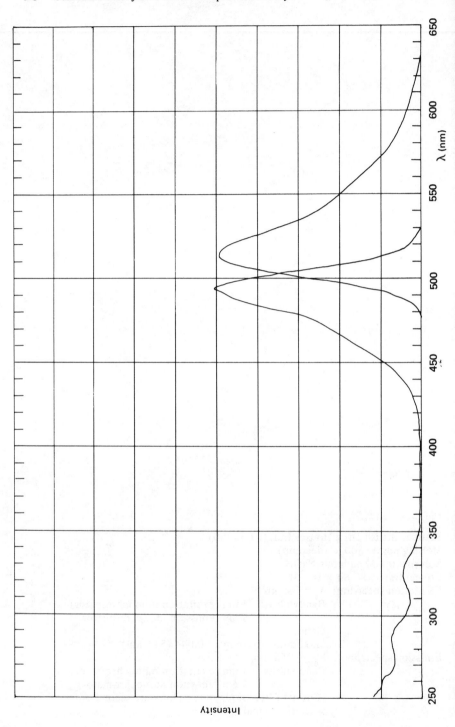

Fig. A.9 FLUORESCEIN
Source: Sigma Chemical Co. Ltd,
Solvent: 0.1 M aqueous NaOH
Concentration: 5×10^{-8} M
Excitation spectrum: λ_f = 520 nm
                      Bandwidths: 3 nm (excitation monochromator)
                                       5 nm (emission monochromator)
                      Gain: 0.1
Emission spectrum: λ_{ex} = 490 nm
                      Bandwidths: 3 nm (excitation monochromator)
                                       5 nm (emission monochromator)
                      Gain: 0.1
(Note that the wavelength scale of this figure is NOT the same as that of the remaining figures)

100 *Standards in fluorescence spectrometry*

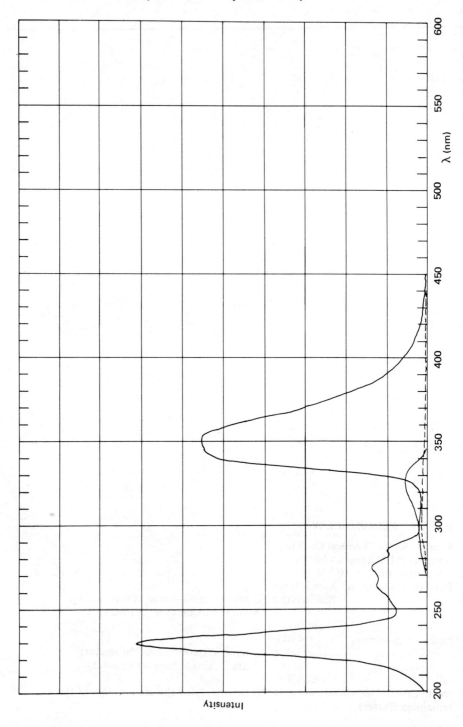

Fig. A.10 β-NAPHTHOL
Source: Sigma Chemical Ltd
Melting point: 122–123°C
Solvent: Tri-distilled ethanol (broken line)
Concentration: 4.63 x 10^{-7} M
Excitation spectrum: λ_f = 335 nm
　　　　　　　　　　Bandwidths: 4 nm (excitation monochromator)
　　　　　　　　　　　　　　　　6 nm (emission monochromator)
　　　　　　　　　　Gain: 1
　　　　　　　　　　Maximum absorbance: 0.035 (226 nm)
Emission spectrum: λ_{ex} = 230 nm
　　　　　　　　　　Bandwidths: 4 nm (excitation monochromator)
　　　　　　　　　　　　　　　　6 nm (emission monochromator)
　　　　　　　　　　Gain: 1

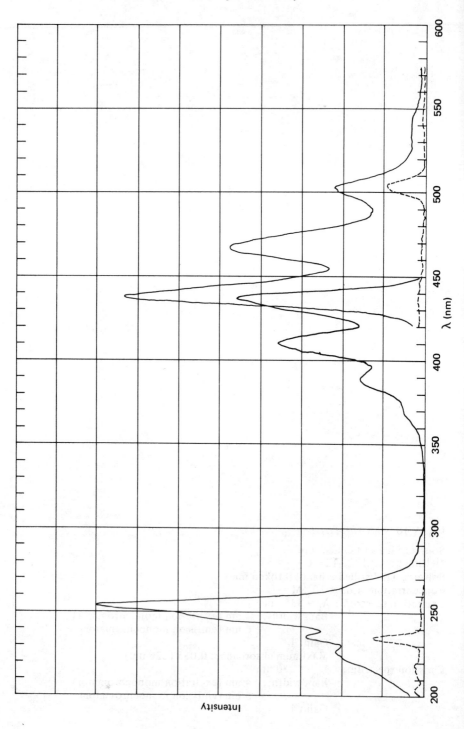

Fig. A.11 PERYLENE

Source: Sigma Chemical Co. Ltd
Melting point: 277–279°C
Solvent: Cyclohexane (broken lines)
Concentration: 3.86×10^{-7} M
Excitation spectrum: λ_f = 465 nm
                      Bandwidths: 4 nm (excitation monochromator)
                                          6 nm (emission monochromator)
                      Gain: 3
                      Maximum absorbance: 0.075 (252 nm)
Emission spectrum: λ_{ex} = 252 nm
                      Bandwidths: 4 nm (excitation monochromator)
                                          6 nm (emission monochromator)
                      Gain: 1

104 Standards in fluorescence spectrometry

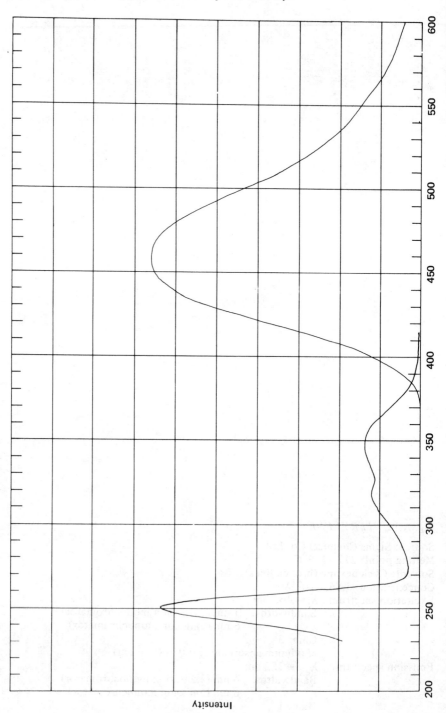

Fig. A.12 QUININE SULPHATE

Source: British Drug Houses Ltd
Melting point: 233-235°C (decomp)
Solvent: 0.105 M aqueous perchloric acid
Concentration: 1.28×10^{-6} M
Excitation spectrum: λ_f = 455 nm
                         Bandwidths: 5.3 nm (both monochromators)
                         Gain: 1
                         Maximum absorbance: 0.07 (250 nm)
Emission spectrum: λ_{ex} = 347 nm
                         Bandwidths: 5.3 nm (both monochromators)

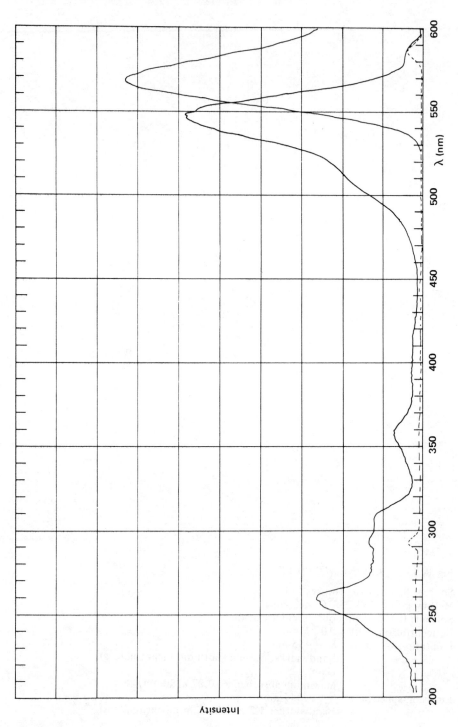

Fig. A.13 RHODAMINE B

Source: British Drug Houses Ltd
Melting point: 210–211°C (decomp)
Solvent: Tri-distilled ethanol
Concentration: 2.8×10^{-7} M
Excitation spectrum: λ_f = 580 nm
                      Bandwidths: 2 nm (excitation monochromator)
                                       12 nm (emission monochromator)
                      Gain: 0.3
                      Maximum absorbance: 0.026 (544 nm)
Emission spectrum: λ_{ex} = 545 nm
                      Bandwidths: 2 nm (excitation monochromator)
                                         6 nm (emission monochromator)
                      Gain: 3

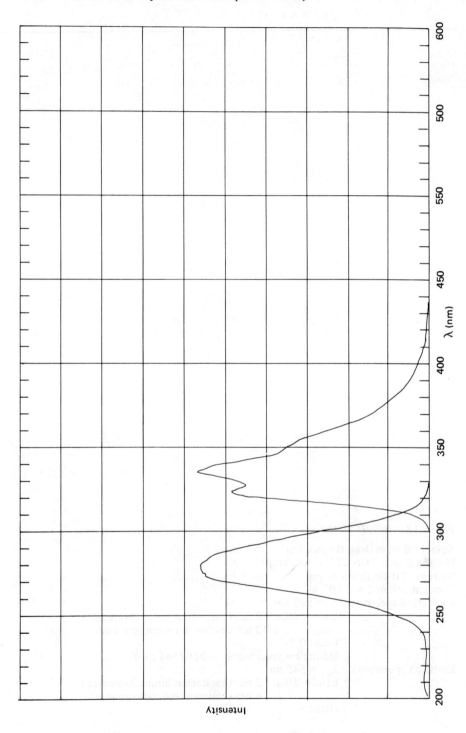

Fig. A.14 *p-TERPHENYL*

Source: Sigma Chemical Co. Ltd
Melting point: 207–209°C
Solvent: Tri-distilled ethanol
Concentration: 1.84×10^{-8} M
Excitation spectrum: $\lambda_f = 335$ nm
                        Bandwidths: 2 nm (excitation monochromator)
                                       4 nm (emission monochromator)
                        Gain: 0.3
                        Maximum absorbance: 0.038 (280 nm)
Emission spectrum: $\lambda_{ex} = 210$ nm
                        Bandwidths: 2 nm (excitation monochromator)
                                       4 nm (emission monochromator)
                        Gain: 0.3

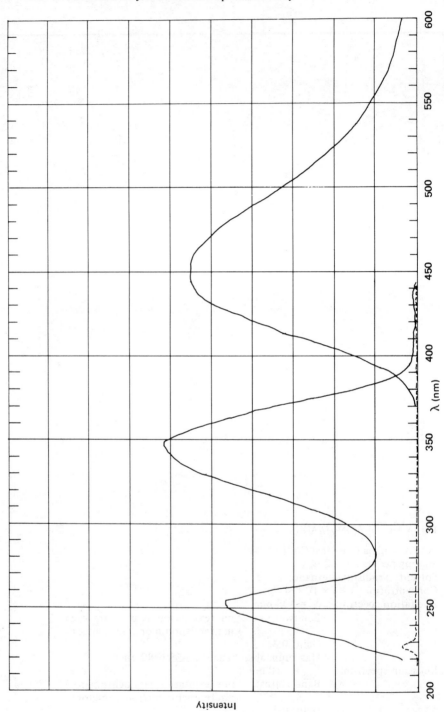

Fig. A.15 *1,1,4,4-TETRAPHENYLBUTADIENE*

Source: Sigma Chemical Co. Ltd
Melting point: 207–209°C
Solvent: Cyclohexane (broken line)
Concentration: 1.2×10^{-6} M
Excitation spectrum: λ_f = 450 nm
                      Bandwidths: 4 nm (excitation monochromator)
                                   6 nm (emission monochromator)
                      Gain: 1
Emission spectrum: λ_{ex} = 345 nm
                      Bandwidths: 4 nm (excitation monochromator)
                                   6 nm (emission monochromator)
                      Gain: 1

112 *Standards in fluorescence spectrometry*

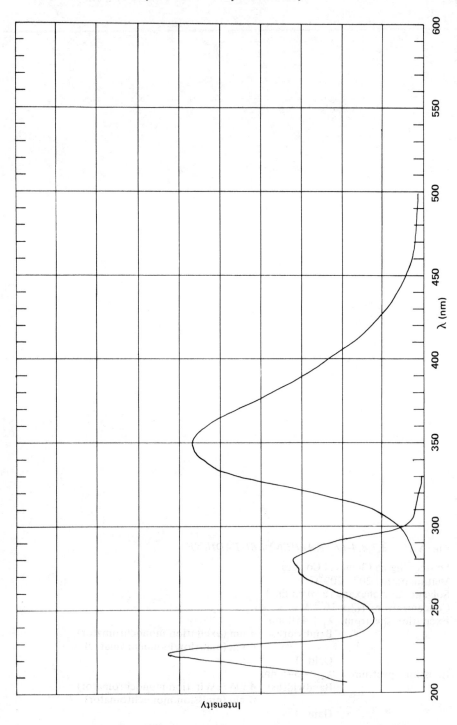

Fig. A.16 *TRYPTOPHAN*

Source: Calibiochem
Solvent: Tri-distilled water
Concentration: 1.1×10^{-6} M
Excitation spectrum: λ_f = 345 nm
                         Bandwidths: 4 nm (excitation monochromator)
                                        12 nm (emission monochromator)
                         Gain: 3
                       Maximum absorbance: 0.03 (218 nm)
Emission spectrum: λ_{ex} = 270 nm
                         Bandwidths: 4 nm (excitation monochromator)
                                        12 nm (emission monochromator)
                         Gain: 3

Index

Acridone, 76
Acrylic copolymers, 13
Actinometer, 47, 52, 73, 74
Aminoacridine, 76
Aminophthalimide, 58, 65
Aminopyridine, 58
Aminopurine, 76
Anisotropy, 57, 60, 62, 64, 70
Anthracene, 14, 16, 59, 61, 82-83, 90-91
Aromatic hydrocarbons, 13
Attenuation factor, 36

Barium sulphate, 65
Beer-Lambert law, 5, 6
Benzene, 69
Benzophenone, 76
Benzoquinoline, 80
Bipyridylosmium complex, 53

Calorimetric methods, 74
Cassegrain optics, 29
Cells, 1
Cerium, 61
Chromium complexes, 76
Colloidal silica, 10
Concentration quenching, 57
Continuous correction, 54
Copper, 61
Coronene, 14, 80, 92-93
Corrected spectra, 1, 4, 8, 49-67, 79-113
Curve corrections, 7
Cyclohexane, 23
Cylindrical cells, 39
Czerny-Turner monochromators, 16

Dansyl sulphonate, 16, 51, 52, 65, 94-95
Delayed fluorescence, 40
Deoxygenation, 42, 47
Depolarizers, 63
Detergents, 5
Dewar flasks, 39, 40

Didymium filter, 12
Dimethylaminonitrostilbene, 76
Dimethylnaphthalene, 69
Diphenylanthracene, 21, 45, 59, 61, 76, 77

Ebert monochromators, 16
Eosin, 59, 71, 84-85, 96-97
Europium, 13, 76
Excimers, 57

Fibre optics, 41
Filters, 4, 9, 24, 55
Flow cells, 39, 47
Fluorescein, 35, 51, 59, 70, 75, 98-99
Fluorescence emission standards, 57-61
Fluorescence intensity, 5-7
Fluorescence lifetime, 2, 28, 44, 75
Fluorescence rate constant, 2
Fluorobenzene, 69
Front-surface geometry, 29-30, 36, 39
Fused quartz, 40
Fused silica, 40, 42

G-factor, 62
Glassy solid standards, 13, 60
Glycogen, 72, 73
Gratings, 9

Haemacytometer, 41
Heavy atom effect, 3, 28, 38
HIDC, 51
Holmium filter, 12
Holographic gratings, 9, 18, 62
Hydroxycoumarin, 76

In-line geometry, 29, 35
Indoleacetic acid, 15
Inner filter effects, 4, 6, 27-38, 41-42, 68-69
Integrating sphere, 74
Interference wedges, 9
Intersystem crossing, 2, 44

Index 115

Lanthanide ions, 13
Laser light sources, 74
Lead, 61
Limit of detection, 15, 20, 21, 22, 24, 39

Magnesium oxide screen, 37, 53, 56, 72
Membrane filtration, 5
Mercury lamps, 4, 12, 16, 24, 69
Mesitylene, 69
Methylene blue, 76
Micro cells, 40
Microcomputers, 7
Microprocessors, 61
Monochromators, 3, 4, 8, 9, 12, 13, 16, 18

Naphthalene, 14, 76
Naphthol, 58, 100-1
Nile Blue A, 51
Nitrodimethylaniline, 58
Nomenclature, 7
Nucleic acids, 13

Ovalene, 14
Oxygen quenching, 3, 28, 57, 59, 60, 71
Ozone, 5

Pathlengths, 36, 37
PBD, 60, 65
Perpendicular geometry, 30, 31, 34
Perylene, 61, 76, 102-3
Phenanthrene, 63, 80
Phenanthroline, 47
Phenylalanine, 69
Phosphorescence, 3, 9, 38
Photodecomposition, 4, 46-48, 75
Photomultiplier, 4, 54
Photon counting, 47, 76
o-Phthalaldehyde, 5
Polarization effects, 54, 57, 62-65, 70, 72, 75, 79
Pontachrome Blue-Black R, 53
Potassium ferrioxalate, 52
Proteins, 13
Pyrene, 61
Pyroelectric detector, 50, 54

Quantum counter, 12, 30, 37, 49-52, 54, 56, 63

Quantum yield, 1, 2, 7, 16, 30, 36, 41, 44-45, 49, 54, 57, 59, 68-77, 79
Quenching, 3, 30, 44, 58, 71, 75, 80
Quinine sulphate, 20-21, 32, 34, 35, 36, 45, 46, 51, 57, 65, 71, 77, 81, 88-89, 104-5

Raman scattering, 3, 21, 22, 23, 24, 25
Rayleigh scattering, 3
Refractive index, 32, 44-45, 69, 70, 73
Reinecke's salt, 53
Rhodamine B, 15, 51-52, 54, 61, 63, 65, 72, 79, 80, 106-7
Rhodamine 101, 51

Sample cells, 5, 39-42, 70
Scattered light, 9, 15, 30, 39, 70
Scrambler plates, 64
Secondary fluorescence, 29, 32
Self-absorption, 27
Sensitivity, 1, 3, 20-26
Signal-to-noise ratio, 20, 23, 24, 25, 46, 80
Spectrometer, 3, 7
Standard lamp, 55, 56
Stern-Volmer law, 32
Stokes shift, 57, 75
Stray light, 1, 4, 15-19
Synchronous scanning, 9

Temperature effects, 4, 44-46, 59, 70
Terbium, 13, 76
p-Terphenyl, 47, 60, 65, 108-9
Tetramethylethylene, 53
Tetraphenylbutadiene, 60, 65, 86-87, 110-11
Thallium, 61
Thermal detectors, 49-50
Thermopiles, 50
Thulium, 13
Tryptophan, 45, 59, 65, 69, 71, 112-13
Tungsten lamps, 55, 79
Tyrosine, 69

Unused orders in gratings, 15

Vavilov's law, 69

White reflectance standard, 56
Windowless cells, 40

Xenon lamps, 4, 10-12, 56, 79, 80